Verner Suomi

Verner Suomi

The Life and Work of the Founder of Satellite Meteorology

John M. Lewis with Jean M. Phillips,
W. Paul Menzel, Thomas H. Vonder Haar, Hans Moosmüller,
Frederick B. House, and Matthew G. Fearon

AMERICAN METEOROLOGICAL SOCIETY

Front cover photograph: Verner E. Suomi (left) and Robert Parent examine radiation balance instrument, 1959 (courtesy of University of Wisconsin Archives).

Published by the American Meteorological Society
45 Beacon Street, Boston, Massachusetts 02108

Print ISBN: 978-1-944970-22-2
eISBN: 978-1-944970-23-9

The mission of the American Meteorological Society is to advance the atmospheric and related sciences, technologies, applications, and services for the benefit of society. Founded in 1919, the AMS has a membership of more than 13,000 and represents the premier scientific and professional society serving the atmospheric and related sciences. Additional information regarding society activities and membership can be found at www.ametsoc.org.

Library of Congress Cataloging-in-Publication Data

Names: Lewis, John M., 1939– author. | Phillips, Jean M., 1958– author.
Title: Verner Suomi : the life and work of the founder of satellite meteorology / by John M. Lewis ; with Jean M. Phillips [and five others].
Description: Boston, Massachusetts : American Meteorological Society, 2018. | Includes bibliographical references and index.
Identifiers: LCCN 2017047857 (print) | LCCN 2017056538 (ebook) | ISBN 9781944970239 (eBook) | ISBN 9781944970222 (pbk.)
Subjects: LCSH: Suomi, Verner, 1915-1995. | Meteorologists-United States—Biography. | Satellite meteorology.
Classification: LCC QC858.S86 (ebook) | LCC QC858.S86 L49 2018 (print) | DDC 551.5092—dc23
LC record available at https://lccn.loc.gov/2017047857

To Anard "Pucky" Suomi
Altruistic Brother of Verner Suomi

Contents

Preface

Verner Suomi's life (1915–1995) began on the Iron Range of northeastern Minnesota, where he displayed extraordinary skills in the mining town's well-equipped industrial arts shops. Work in a depression-era Civilian Conservation Corps camp led to an opportunity to attend Winona State Teachers College, where his studies in education spared him from a life in the mines. Suomi then assumed a settled life as a schoolteacher of science on the prairies of south-central Minnesota for four years. But the vicissitudes of World War II put him on another path—the "second" life. And although this second life started innocuously as a meteorological trainee in the University of Chicago's Cadet Program, he steadily advanced in academia as a researcher and teacher and then as a professor at the University of Wisconsin–Madison, where he founded the Space Science and Engineering Center (SSEC), came to be known as "the father of satellite meteorology," and was awarded the National Medal of Science in 1977.

In this scientific biography, the reader will see that Suomi was that rare combination of "unifier" and "diversifier" in Freeman Dyson's definition of physicists in his stimulating book *Infinite in All Directions* (Dyson 1988). Suomi chose his problems with the simplicity and majesty of an abstract thinker, but he never overlooked a detail, especially in the construction of a "gadget" to make measurements on the path to solving a problem that held promise for benefiting the world's citizenry. These qualities are praised in a passage from a letter written by future director of the SSEC Hank Revercomb to Suomi on July 20, 1995 (10 days before his death):

> You engender the super energy and power that can be unleashed by genuinely inspired interest in solving important problems for mankind (and the challenge of doing it).

Suomi defined variance of the human spirit. He was like a crystal in the sunlight—exhibiting a spectrum of colors. If you only knew his name, it is likely that the refracted ray would spell "satellite meteorology." But if you knew him from a conference or seminar, you would remember him for his simplicity of expression on a complicated topic, ease of talking with animated gestures, and vocalizing with a melodic tone. And if you worked at SSEC or otherwise knew him as a colleague, you would remember him as a friend, if an outspoken one.

This biography follows Suomi's path, far from orderly, to prominence, revealing an almost mystical fortuity: his creative energy perfectly coincides with the spirit of the time, his natural talent with the world's mania for space and space science. Without planning, he had prepared himself perfectly for the marriage between the artificial satellite and scientific measurements from space. And with the good fortune of striking friendship and collaboration with Robert Parent, the brilliant electrical engineering professor at UW–Madison, together they succeeded in measuring the Earth's heat budget with instrumentation aboard *Explorer VII*.

This scientific biography of Suomi was inspired by events in 2009 surrounding the 50-year celebration of the *Explorer VII* mission. Following the celebration, key organizers Jean Phillips and Paul Menzel were inspired to initiate a Suomi biography project. It would be an elaboration and expansion on the article written by John Lewis and coauthors, "Suomi: Pragmatic Visionary" (Lewis et al. 2010). Thus, in 2012, a team consisting of Suomi colleagues and protégés was formed to write the biography. Rather than a set of separate articles that describe the character and work of Suomi, a decision was made to construct a continuous storyline with input from all authors and a host of other contributors. We have strived to present this story in a singular voice that reflects our collective passion and knowledge.

Acknowledgments

We have benefitted greatly from interviews with Suomi's children (Lois Young and Eric Suomi) and his niece (Judi Suomi Maki), with Robert Parent's children (Patti Calloway and Barbara Parent), and with Suomi's students, protégés, and colleagues, many still active in science. Interviews with the following scientists were especially valuable:

Eric Anderssen
William Bourke
Stephen Cox
John Eyre
Robert Fox
Douglas Gauntlett
Thomas Haig
Martin Hoerling
Donald Johnson
Graeme Kelly
John LeMarshall
Ronald McPherson
Graham Mills
Pierre Morel
Henry Revercomb
James Rasmussen
Douglas Sargeant
Johannes Schmetz
William Smith
Lawrence Sromovsky
John Young
Christopher Velden

Others, whose oral histories and/or personal communications added details to the scientific biography, are:

Douglas Allen
Francis Ashley
Werner Baum
Reid Bryson
Horace Byers
Jule Charney
Dave Fultz
Kenneth Johnson
Julius London
Pierre Morel
Noburo Nakamura
Donald Osterbrock
Norman Phillips
George Platzman
Carl Rossby
Timothy Schmit
Henry Stommel
James Weinman
Aksel Wiin-Nielsen
William Togstad
Harriet Woodcock
Oliver Wulf

We also benefitted from written tributes to Suomi by:

Fred Best
Eugene Buchholtz
Joost Businger
William Hibbard
David Jones
Scott Lindstrom
John Roberts
Henry Schmit
Charles Stearns
Anthony Wendricks

Archivist Richard Popp scoured the Announcements in the Physical Sciences at the University of Chicago and uncovered the requirements for the PhD in meteorology, and archivist Elizabeth Andrews at MIT delivered important materials that helped define Rossby's view of mentorship.

The Suomi Archive at the University of Wisconsin–Madison's Schwerdtfeger Library includes hundreds of formal letters and informal notes, sketches of his engineering ideas, photographs, interviews (transcribed and audible), and a complete set of his technical papers, formal publications, and patents. It provided the solid underpinning for the project. A special thanks goes to Jean Phillips, the UW–Madison archivist and librarian with an encyclopedic knowledge of Verner Suomi's life, who was the stimulus behind the project and coauthor of the book, as well as to her colleague, Linda Hedges.

We are indebted to Bill Bourke for an invaluable review of the manuscript. As typical of good reviewers, Bill identified references that added substance to various chapters, suggested changes in wording that clarified arguments, and asked penetrating questions that led to some restructuring of the manuscript.

Finally, we thank the Publications Department of the American Meteorological Society (AMS) for sponsorship and for offering valued advice at every phase of the four-year project. Those deserving special credit are the editors, Sarah Jane Shangraw (AMS Books Managing Editor) and James Fleming (AMS Books History Editor at Large). Their encouragement gave us energy and purpose. They, in turn, were both strongly supported by Ken Heideman (Director of Publications) and Beth Dayton (AMS Books Production Manager), among other excellent Publications staff.

I

The Winding Road to Meteorology

CHAPTER ONE

On the Mesabi during the Great Depression

Verner E. Suomi (1915–1995) was born and raised in Eveleth, Minnesota, the sixth of seven children—five girls and two boys.[1] In his unique way of viewing life, he commented, “I always complained that my sisters had two brothers and I only had one” (Suomi et al. 1994). And, indeed, Vern’s brother Anard, nine years his elder, became Vern’s guiding light [see Vignettes, Judi (Suomi) Maki].

Eveleth was one of the prominent mining towns on the Mesabi Range. Much of the world’s supply of iron ore came from the cavernous open-pit mines along this twisting, narrow, 120-mile stretch of land in northeastern Minnesota (Figure 1.1). Industrial giants clamored for the rights to this land after the high-grade ore (hematite) was discovered in 1892, and by the end of the century the productive parts of the range were identified. There was an immediate need for strong men to work the mines, and immigrants from across the Atlantic descended on the towns of Grand Rapids, Hibbing, Virginia, Eveleth, Biwabik, and Babbitt, the hotbeds of iron ore along the Mesabi. Vern’s mother and father emigrated from Finland to Eveleth in 1902, where Vern’s father found work as a carpenter in the mines.

1. Verner Suomi’s parents were John E. and Anna Emelia (née Sundquist) Suomi.

Figure 1.1. North-central states map that chronologically tracks cities where Verner Suomi lived and worked (courtesy of Sarah Witman).

The land was raw with outcroppings of ancient bedrock, overlain in places by layers of gravelly soil left by the advance and retreat of glaciers during the ice ages. Extraction of the iron ore, oft times imbedded in boulders along angled ribbons of rock beneath the soil, proved challenging for both steam-driven shovels and men with pick, shovel, and wheelbarrow who toiled beside the machinery. Pictures of the deep open-pit Spruce Mine in Eveleth and the men who worked that mine are shown in Figures 1.2 and 1.3. Eveleth was nudged up against this mine, and an aerial photograph of the city is shown in Figure 1.4.

Suomi's oral history interviews (Suomi and Lewis 1990; Suomi et al. 1994) reveal that the family was not poverty-stricken. Nevertheless, conditions were only a notch or two above the hardscrabble life—conditions that included a

Figure 1.2. Spruce Mine, an open-pit iron ore mine on the outskirts of Eveleth, MN, ca. 1930 (courtesy of the Minnesota Historical Society).

Figure 1.3. Miners who worked at the Spruce Mine, ca. 1930 (courtesy of the Minnesota Historical Society).

Figure 1.4. Aerial view of Eveleth, MN, ca. 1940 (courtesy of Minnesota Historical Society).

house and a bunkhouse (where Anard and Vern resided in spring, summer, fall, and early winter), both structures without the benefits of running water and electricity. Their food primarily came from a large garden, and the family canned the excess to provide a year-long supply of vegetables. The children had limited but adequate clothing, and, in his characteristically optimistic manner, Vern said,

> it was basically a happy family. We were Lutherans and went to church every Sunday, and then Sunday School . . . my mother was a remarkable woman [with severe arthritis][2]. . . . And despite the fact that she was bedridden, she had good control of the family, and of course that gave much of the work to my sisters. . . . This was during the Depression, of course, when there weren't any jobs [for women in Eveleth] . . . so each of the sisters would gradually go away and become a domestic somewhere in Chicago. . . . My father worked so hard, ten hours a day, so we didn't do too many things together. He was just exhausted. (Suomi et al. 1994)

Using discarded thread spools for wheels, Vern built his own toys—typically replicas of steam shovels and other earth-moving equipment that

2. Information inserted by the authors has been bracketed.

Figure 1.5. A "sticker" with the Finnish word *sisu* overlaying the Finnish flag seen on the bumpers of cars and trucks in the Upper Peninsula of Michigan and the Iron Range cities of Minnesota. *Sisu* is a Finnish word which loosely means stoic determination, grit, bravery, resilience, and hardiness and is held by Finns themselves to express their national character.

he saw in the mines. Vern's recollection of his introduction to the operation of a wristwatch follows:

> As you might imagine, I was excited when my sister got a wristwatch for Christmas. After she went to bed that night, I took the watch and disassembled it. I couldn't get it back together even though I stayed up most of the night. She knew who broke it and my punishment was severe! (Suomi 1985, personal communication)

Judi Maki, Anard's daughter, remembers her father saying, "Vern always took apart more than he put together" [see Vignettes, Judi (Suomi) Maki].

In school, Vern took part in debate and joined the aviation and radio clubs. His greatest joy came from the excellent industrial arts courses he took at Eveleth High School. The large L-shaped building to the right of center in Figure 1.4 was the home of the city's high school and junior college. In Vern's words,

> the schools were absolutely superb. You see, with the mines there, there was a tax called the *ad valorem* tax [proportion to the value] which gave even a little city like ours, which was only 9,000 people, a remarkable budget. (So we had machine shops and glass shops and printing shops, woodworking

> shops, auto mechanic shop . . . I took these courses where you learn to use your hands . . . and you've got to get things done . . . finish it! I kept that with me throughout my life as a scientist). (Suomi et al. 1994; Suomi and Lewis 1990, in parentheses)

The Finnish word *sisu* describes this trait of "finishing a job"—taking action against the odds and displaying resoluteness in the face of adversity. It is widely considered to lack a proper translation into any other language (Schatz 2005). When travelling through northern Minnesota and the Upper Peninsula of Michigan, you will occasionally see a bumper sticker with the word *sisu* printed over the Finnish flag. One of these bumper stickers is shown in Figure 1.5.

CHAPTER TWO

Road to Winona and Chicago

Although Vern showed promise in industrial arts and debate, his future was uncertain upon graduation from Eveleth High School in 1933. The family's economic state suffered after Vern's father lost his job in 1926. His father did not qualify for a pension since he fell just short of the required 25 years of employment in the mines. Anard dropped out of his first year at Eveleth Junior College and became the family's breadwinner as a laborer in the mines (J. Maki 2016, personal communication). In his oral history interview (Suomi et al. 1994), Vern remembered the situation with a choked-up tone in his voice: "He [Anard] was unable to complete school because of financial problems, but he would have been a great one if he did. . . . I said to him, 'Why don't you go out for yourself?' and he said, 'You're supposed to honor your father and mother.' And those things have stuck with me for a long, long time."

Taking the role of breadwinner to heart, Anard was able to save enough money to send Vern to Eveleth Junior College. The tuition was $14 per semester. As Vern remembered, "[Anard said] I will put you through college, but you'll have to earn the first $14" (Suomi et al. 1994). And, indeed, Vern earned that first $14 by working at every conceivable job during that summer of 1933. Anard, good to his word, supported Vern through two years of junior college, where he took courses in mathematics and science.

Upon graduation from junior college, Vern was able to find a job with the Civilian Conservation Corps (CCC) at Camp S-52 in Orr, Minnesota, about 50 miles north of Eveleth (the camp is described in Sommer 2008). He worked as a handyman, using his industrial arts skills to make mechanical and electrical repairs at the camp. While there, he came under the influence of the camp's educational advisor, John Blotnick,[1] who had connections to Winona Teachers College in southeastern Minnesota and secured a scholarship for Vern at the college (Figure 1.1). It was a partial scholarship; another $200 per academic year would be needed. And once again, Anard managed to dig up that money in $20 per month increments.

Vern entered Winona Teachers College in the fall of 1935. He said, "I don't recall that I was a superstar pupil, I wasn't a dumb one either" (Suomi et al. 1994). He helped defray expenses at Winona by working as a laboratory assistant in Professor Nels Minne's chemistry class.[2] Before completing the course, he left school for one semester during the fall/winter of 1936/37, to work and save money for his final year of undergraduate education. As he remembered,

> I left school for one fall and one winter and I took a civil service exam for the lowest job I ever had, which was junior assistant technician for the U.S. Forest Service. They don't come any lower. I worked at a CCC camp where the Army took care of our living aspects when you came in camp at nights, whereas in the morning after reveille, you were turned over to the Forest Service. When I came back to school I was in class plays. I liked that and that's where I met Paula [Vern's future wife].[3] (Suomi et al. 1994)

Although Vern never aspired to be a teacher, the bachelor of education (B. E.) degree, granted in 1938, set him off on a career as a junior high school and high school science teacher in south-central Minnesota. He taught for four years in three different cities [New York Mills, Sleepy Eye, and Faribault (Figure 1.1)]. While at Faribault, he received his pilot's license as part of the Civilian Aviation Authority's (CAA) program. The ground school in the CAA program gave Suomi his first taste of meteorology. He remembered,

---

1. Blotnick later served as U.S. congressman from Minnesota's 8th District (essentially the Iron Range) for seven consecutive terms (1946–1974).

2. Minne later served as President of Winona Teachers College for a period of nearly 25 years (1944–1967).

3. Paula (née Meyer) and Vern were married on 10 August 1941.

"I really liked that little book about meteorology, a book about aviation meteorology by Benarthur Hayes" (Suomi and Lewis 1990).

With the likelihood of the United States' entry into World War II (WWII) by late 1940 or early 1941, Vern had the draft board "breathing down my neck" (Suomi et al. 1994). Then he heard a half-hour Mutual Radio Network broadcast where Carl-Gustaf Rossby, professor at the University of Chicago, announced a 9-month Cadet Program for college graduates who were interested in becoming meteorologists (Rabson 1998). This program was aimed at increasing the number of weather forecasters in the military and civilian communities. The phrase "Victory through air power" caught Suomi's attention (Suomi and Lewis 1990). There were two avenues for entry into the Cadet Program: 1) a military path where graduates would be given commissions in the Army Air Force (AAF) or the United States Navy (USN) and 2) a civilian path open to students who had completed the CAA's pilot training course.

Vern applied and indicated he would be willing to enter the program through either avenue. He received a Western Union telegram from Rossby. In the telegram, Rossby asked Suomi to consider entry into the Cadet Program via the civilian path since there were fewer applicants on that side. The pay was less for students in the CAA program, but Suomi honored Rossby's request and enrolled in the third cadet class at the University of Chicago that began on 1 March 1942. Civilians received $75 per month (without tuition coverage), whereas the military students received $140–$145 per month with tuition coverage (Cadet Program 1941). There were 125 students in Suomi's class—104 military students and 21 civilian students. The Cadet Program trained over 7,000 students at six institutions between October 1940 and June 1944 (Walters 1952; Byers 1970). In addition to the University of Chicago, the other institutions that offered a Cadet Program included the Massachusetts Institute of Technology (MIT); California Institute of Technology (Caltech); University of California, Los Angeles (UCLA); New York University (NYU); and the Army Air Force Technical Training Center, where classes were held in the Pantlind Hotel in downtown Grand Rapids, Michigan (later moved to Chanute Field, Illinois).

CHAPTER THREE

Entrée into Meteorology: Cadet to Instructor at the University of Chicago

Suomi had little attraction toward meteorology as a career path. He viewed his assignment to the Cadet Program as follows: "I was interested in becoming an electrical engineer . . . I was not interested in being a meteorologist. I was mainly interested in not getting into the infantry" (Suomi et al. 1994). The courses were challenging, with a raft of excellent professors including Rossby, Horace Byers, Victor Paul (V. P.) Starr, and Michael Ference (professors and instructors are listed in Figure 3.1 and courses are listed in Figure 3.2). The outline of material covered in Dynamic Meteorology 315 is displayed in Figure 3.3. Considering that this material was covered in an 11-week period, it was no mean feat for a Cadet to excel.[1] The competition included some excellent students who already possessed graduate degrees in scientific fields other than meteorology (R. Bryson 1990, personal communication).

The first-quarter grades for the University of Chicago's second cadet class (July 1941–February 1942, the class before Suomi was enrolled) are shown in Table 3.1. The distribution is certainly not skewed toward high grades. Vern received a "C" in Dynamic Meteorology 315 (Figure 3.3). As he recalled,

---

1. The quarter consisted of 12 weeks, 1 week for orientation and 11 weeks of instruction (Rossby 1943).

INSTITUTE OF METEOROLOGY
OFFICERS OF INSTRUCTION

Carl-Gustaf Arvid Rossby, Sc.D., Director of the Institute of Meteorology and Andrew MacLeish Distinguished Service Professor Meteorology.
Horace Robert Byers, Sc.D., Secretary of the Institute and Professor of Meteorology.
Helmut Erich Landsberg, Ph.D., Associate Professor Meteorology.
Erwin Reinhold Biel, Ph.D., Visiting Associate Professor Climatology.
Michael Ference, Jr., Ph.D., Associate Professor Meteorology.
Victor P. Starr, Sc.M., Assistant Professor of Meteorology.
Phil Edwards Church, Ph.D., Research Associate in Oceanography.
Oliver Reynolds Wulf, Ph.D., Research Associate in Physics.
Vincent J. Oliver, S.B., Instructor in Meteorology.
Lynn L. Means, S.B., Instructor in Meteorology.
Herbert Riehl, S.M., Instructor in Meteorology.
Leonid Hurwicz, LL.M., Research Associate and Instructor in Meteorology.
George W. Platzman, S.M., Instructor in Meteorology.
Mildred Boyden Oliver, A.B., Instructor in Meteorology.
John Cary Bellamy, Ph.M., Instructor in Meteorology.
Clayborn D. McDowell, S.B., Instructor in Meteorology.
Sidney Teweles, S.M., Instructor in Meteorology.
George T. Benton, Instructor in Meteorology.
Glenn E. Stout, Jr., A.B., Instructor in Meteorology.
Robert Gardner Beebe, A.B., Instructor in Meteorology.
David Hinshaw Shideler, S.B., Instructor in Meteorology.
Robert Richard Bentley, S.B., Instructor in Meteorology.
Albert Cahn, Jr., A.B., Instructor in Meteorology.
Jack Indritz, S.M., Instructor in Meteorology.
Lawrence F. Markus, S.B., Instructor in Meteorology.
George G. Cohn, A.B., Instructor in Meteorology.
Josephine Ann Peet, S.B., Instructor in Meteorology.
Abe Rosenbloom, S.M., Instructor in Meteorology.
John G. Phillips, S.B., Instructor in Meteorology.
Verner E. Suomi, B.E., Instructor in Meteorology.
Eleanor Martha Hanson, S.B., Instructor in Meteorology.
Ward Chennell, Assistant in Meteorology.

INTRODUCTORY

The Institute of Meteorology has as its objectives: (1) advancement of the understanding of atmospheric processes and measurements, (2) instruction in the principles of meteorology, and (3) training of graduate students for professional work in meteorology. To meet the need for great numbers of newly trained professional meteorologists in the military and civil weather services, the Institute has geared its instruction to a highly intensified, productive program. A majority of the students are officers and men of the United States Army Air Forces. In addition to its all-out war instruction program, the Institute of Meteorology has taken a forward position in research directed toward the solution of problems of foremost importance in both war and peace.

FACILITIES

For meteorological instruction and research, the University provides complete synoptic-analysis laboratories, including files of several years of analyzed weather maps and upper-air charts for various regions of the world. The usual meteorological instruments and equipment for pilot-balloon observations are used by the students, and a completely equipped laboratory for radiosonde observations of the upper atmosphere is provided. A special feature is the mobile weather unit which, mounted on a truck, provides experience similar to that encountered in combat areas. The mobile unit is equipped with pilot-balloon and radiosonde apparatus, and weather-charting equipment permitting a complete weather observation, analysis, and forecast in the field. A complete hydrodynamics laboratory is used in the instruction and research. An official observatory of the United States Weather Bureau is located on the Quadrangles. Well-equipped shops make possible construction of special apparatus for research. A special library of meteorological books and journals is in the Institute of Meteorology.

Figure 3.1. Officers of Instruction, Institute of Meteorology, University of Chicago, Academic Year 1943/44.

COURSES OF INSTRUCTION

For information concerning quarters when courses are given, see "Schedule of Courses" above. Courses 312 to 319, inclusive, offer credit either in Meteorology or in Physics. Courses marked * require payment of a laboratory fee (see p.21).

201. Introductory Meteorology.--Structure of the atmosphere; atmospheric motions and meteorological processes; air masses and fronts; tropical and extratropical cyclones. Tu-F, 10, BYERS, STARR.

203. Applied Climatology.--The climates of the world with special emphasis on war theaters. Tu-F, 9, LANDSBERG, BIEL.

205. Field Course.--Instruments and technique of weather observations; special atmospheric measurements at the observatory and in the field. ½C. Hrs to be arranged, LANDSBERG, STAFF.

211. Synoptic Meteorology.--Radiation, convection, evaporation, etc., in relation to the properties of air masses; formation and structure of fronts; the tropical and extra-tropical cyclones. Tu-F, 11, BYERS, RIEHL.

212. Synoptic Meteorology.--Details of frontal activity; motions on isentropic surfaces; lateral divergence and vorticity; physics of condensation and precipitation; phenomena affecting aeronautics. Tu-F, 11, BYERS.

***216. Meteorological Laboratory.**--Synoptic weather observations; decoding and plotting of synoptic messages; isobaric and frontal analysis of the weather map and practice in the utilization of upper-air data. 2Cs. Afternoons and evenings, MEANS.

***217. Meteorological Laboratory.**--Three-dimensional synoptic analysis, including isentropic and other upper-air charts; displacement of pressure systems and fronts; forecasting practice. 2Cs. Afternoons and evenings, OLIVER, MEANS.

***218. Meteorological Laboratory.**--Complete three-dimensional synoptic analysis, including isentropic and other upper-air charts; displacement of pressure systems and fronts; forecasting practice. 2Cs. Afternoons and evenings, OLIVER.

***221. Oceanography.**--The physical geography of the sea; elementary dynamics of ocean currents ½C. Tu-F, 9, CHURCH.

***246. Advanced Calculus for Meteorologists.**--Selected topics in advanced calculus, including partial derivatives, line and surface integrals, and the elements of ordinary and partial differential equations with emphasis on applications to mechanics and physics. [Not given 1943-44].

312. Hydrodynamics.--Review of classical mechanics including equations of motion in rotating coordinate systems. Hydrodynamics of nonviscuous fluids, Euler's equation, potential flow, vortex motion. Concepts illustrated with experiments. Tu-F, 8, FERENCE.

313. Hydrodynamics of Viscous Flow.--Continuation of classical hydrodynamics with emphasis on special problems in potential flow. Navier-Stokes' equation, creeping motion, applications to lubrication, introduction to boundary layer theory. This course includes a few laboratory experiments. Tu-F, 10, FERENCE.

314. Hydrodynamics of Turbulent Flow.--Special topics in turbulent motion, including laboratory work. Hrs to be arranged, FERENCE.

315. Dynamic Meteorology.--The general circulation of the atmosphere; thermodynamics and statistics. Prereq: Physics 105, 106, 107, and preferably Math 247. Students without Met 312 are urged to take it simultaneously. Tu-F, 11, ROSSBY, STARR, PLATZMAN.

316. Dynamic Meteorology.--Introduction to the dynamics of the atmosphere. Prereq: Met 315, 312. Tu-F, 8, ROSSBY, STARR, BELLAMY.

317. Dynamic Meteorology.--Application of dynamic concepts to forecasting and special problems related to warfare. Tu-F, 8, ROSSBY, BELLAMY.

318. Physics of the High Atmosphere.--Composition and properties of the atmosphere at great heights; atmospheric ozone; radiation and ionization. Tu-F, 10, WULF.

***319 Technique of Upper-Air Observations (laboratory).**--The various types of radiosonde and other instruments for observations in the upper air. ½C. Hrs to be arranged, BELLAMY.

341, 342, 342, 344. Experimental Meteorology.--Introductory courses leading usually to a Master's thesis. Problem will be assigned in consultation with the instructor. Prereq: Met 216, 217, 218, 319. Hrs to be arranged, ROSSBY, BYERS, LANDSBERG, FERENCE, STARR.

399. Professional Forecasters' Course.--A special, intensive course for professional meteorologists, involving mainly individual study of the application of results of new research to weather forecasting. Registration only after consultation with appropriate instructor. 3Cs. Each qr, hrs to be arranged, STAFF.

406, 407, 408, 409. Research Course.--For students prepared to undertake special research, leading to a Doctor's thesis. Hrs to be arranged, ROSSBY, BYERS.

420. Reading and Research in Foundations of Meteorology.--Prereq: Facility in reading German. Each qr, STAFF.

* Optional.
† May be taken first quarter.
‡ Geography 203 and Meteorology 221 are two half-courses in sequence.

Figure 3.2. Courses of Instruction, Institute of Meteorology, University of Chicago, Academic Year 1943/44.

Dynamic Meteorology - Starr (instructor)

Course outline 1st Quarter

I General circulation
II Gas Laws & Humidity measures
III Adiabatic processes & Pot. temp. (Entropy)
IV Adiabatic chart & pressure height relations
V Moist adiabatic processes
VI Thermodynamic charts in general

Reference books

1. Scientific basis of Meteorology U of C - Rossby
buy * 2. Dynamic Meteorology MIT. - B. Haurwitz
3. Weather analysis & forcasting - S. Petterssen
* 4. Physical & Dynamic Met. - D. Brunt
5. Thermodynamics - M. Planck
6. Manual of Met. English - N. Shaw
7. Physics of the air - W.J. Humphreys
8. Dynamische Met German - F Exner
9. Synoptic & Aeronautical Met. U of C - H. Byers

Exam. Apr. 15

derive [illegible] relations...

John – this is a copy of the class notes I took in Starr's 1st Quarter class starting in March 1942 in the cadet Met. course at the U. of Chicago
– Ken

Figure 3.3. Outline of material covered in Dynamic Meteorology 315, Instructor V. P. Starr, Suomi's Cadet Class of 1943 (courtesy of Ken Johnson).

Table 3.1. Grades for second cadet class at the University of Chicago (July 1941–February 1942). GPA = grade point average and C = course credit.

Course	Army Air Corps Officers (6)	Air Corps Cadets (22)	CAA Graduates (22)	GPA
Intro to Meteorology Meteor 201: 1C	4C, D, F	5A, 8B, 3C, 6D	4A, 3B, 9C, 6D	2.28
Meteorology Laboratory Meteor 216: 2C	B, 5C	3A, 11 B, 7C, D	A, 9B, 11C, D	2.54
Meteorology Field Course Meteor 205: ½C	2B, 3C, D	2A, 16B, 4C	4A, 17B, C	2.48
Dynamic Meteorology Meteor 315: 1C	C, 4D, F	2A, 5B, 9C, 6D	2A, 8B, 5C, 6D, F	2.02
Upper-air Observations Meteor 319 ½C	2B, 4C	5A, 12B, 5C	4A, 5B	3.00
GPA	1.83	2.66	2.59	

> I didn't even have a course in differential equations when I came to Chicago. (I discussed my disappointing performance with Professor Ference, who taught the course. He encouraged me and I got an "A" in the next dynamics course [Dynamic Meteorology 316] . . . he was the best teacher I ever had in my life.[2]) (Suomi et al. 1994; Suomi and Lewis 1990, in parentheses)

Professor Ference is pictured in Figure 3.4.

One of Vern's memorable experiences at the University of Chicago was instruction under Tor Bergeron, the noted Bergen School synoptician. Bergeron visited the university and gave lectures in the synoptic meteorology courses (Synoptic Meteorology 211, 212). In recollection, Vern said,

2. Following receipt of his PhD in experimental physics at the University of Chicago in 1936, Ference was appointed professor of physics (and later meteorology) at the university. He left the university after WWII to eventually become vice president in charge of research for Ford Motor Co., and later served as presidential advisor to both Lyndon B. Johnson and Richard M. Nixon. He was elected to the National Academy of Engineering in 1971 (Harwood 2001).

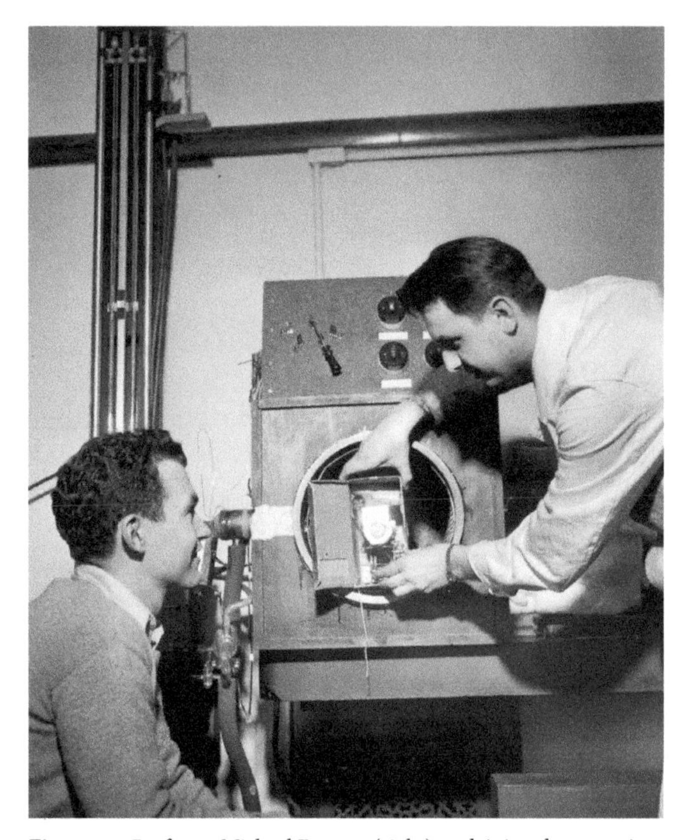

Figure 3.4. Professor Michael Ference (right) explaining the operation of a radiosonde to a student in the University of Chicago's Cadet Program. A pressure chamber is used to calibrate the radiosonde's anemometer (courtesy of the University of Chicago Archives).

> You see, from the front porch of his house, he took all these pictures of the weather. The mountains were in the background and changes in visibility were obvious, and the clouds. He could explain the changes in the weather—winds, clouds, and visibility—in such an interesting way. (Suomi and Lewis 1990)

And then Rossby had an approach to teaching that thrilled Vern. As he remembered,

> Rossby always started with the physics, then he'd demonstrate the solutions graphically, and finally give us the mathematics. His intuition was phenomenal and he fooled a lot of the sharp mathematical students in the back row! (Suomi and Lewis 1990)

Rossby's own view of his teaching was expressed clearly and succinctly in a wide-ranging letter he wrote to Jule Charney in 1952:

> Perhaps I occasionally sought to give, or inadvertently gave to the student a sense of battle on the intellectual field. If all you do is give them a faultless and complete and uninhabited architectural masterpiece, then you do not help them to become builders of their own. (Rossby 1952)

While not a student of Rossby, Werner Baum, another member of the graduate student corps at the University of Chicago in the late 1940s, had similar sentiments about Rossby's overall influence on the students:

> As you [J. L.] probably know, I was not a student of Rossby's . . . [but] in another sense, of course, everyone there [at the University of Chicago] was Rossby's student, because his presence was pervasive. The interminable parade of the leading meteorological personalities of the world, Rossby's complete devotion to the search for understanding of the atmosphere and oceans, his treatment of a graduate student as anyone's equal if the student displayed a good mind, his intuitive ability to discern the correct from the incorrect (to the extent of making canceling mathematical mistakes and arriving at the correct answer), and his charm (which certainly would have been a great asset in the diplomatic service) infected everyone. He set an example of scholarly excellence and diligence which followed us all our professional lives. (W. Baum 1991, personal communication)

Photographs of Rossby are found in Figure 3.5.

As a cadet, Vern offered assistance to Professor Ference, who was in charge of the radiosonde lab at the University of Chicago. The balloon-borne upper-air "soundings" (temperature, pressure, and water vapor observations) were transmitted by radio and labeled "radiosondes." These instruments were in short supply, and farmers and others returned many damaged sondes to the university. With Vern's skill in engineering that followed him from Eveleth, he became the "master mechanic of radiosonde repair" (Suomi and Lewis 1990).

Graduation for the third class of cadets at the University of Chicago took place on 30 November 1942. The photograph taken on graduation day is shown in Figure 3.6. Vern must have moved when the photographer took this picture since his image is blurred (upper left-hand corner of the photograph). Figure 3.7, supplied by Reid Bryson (1990, personal communication,

Figure 3.5. Carl Rossby examining an orchid on the island of Gotland (his signature beneath the photo), and the inset shows Rossby viewing a hydrodynamics experiment in Dave Fultz's Hydro Lab at the University of Chicago (courtesy of H. Thomas Rossby, Dave Fultz).

Geophysical Sciences Department, University of Chicago), displays a chart that identifies students and teachers in this photograph.

By the end of the 9-month program, Suomi realized there might be a place for him in meteorology. He was not attracted to weather forecasting, but rather to weather observing. And because of his obvious skill with instrumentation and strong performance in the academic courses, Horace

Figure 3.6. Graduation photograph of the University of Chicago's third Cadet Class (courtesy of Noburo Nakamura and the Department of Geophysical Sciences, University of Chicago).

APPENDIX A. A refined version of line drawings of the heads that correspond to Fig. 3 (Original Courtesy of the Department of Geophysical Sciences, U of C). Reid Bryson identified the following faculty and students in Fig. 3: Verner Suomi (16); Reuben Belongia (41); Reid Bryson (45); Ben Bullock (49); Lynn Means (52); Oscar Singer (54); Robert Beebe (58); Ralph Nelson (59); John ? (62); Fred White (63); Bill Plumley (65); Micheal Ference (66); John Finch (67); Chidley [Ken] Johnston (69); Larry Curtis (72); John X. Jamrich (75); Mike Chancellor (80); Glenn Stout (81); Phil Smith (82); Earl Fowler (83); Vince Oliver (98); W.T. Reid (101); Phil Church (102); Herbert Riehl (103); Victor Starr (104); Leonid Hurwicz (105); Raymond Wexler (106); Joshua Holland (108); Lawrence Markus (110); Robert Bentley (112); George Haltiner (113); George Platzman (114); Massey (115); Jack Indritz (116); ? Stanley Beloy (118); Frank Snodgrass (121); Earnest Bice (122); Oliver Wulf (123); Capt. Starbuck (124); Carl Rossby (125); Horace Byers (126).

Figure 3.7. Identification of faculty and students via outline drawings of individuals in Figure 3.6.

Byers, secretary of the Institute of Meteorology, offered him a position as an instructor at the University of Chicago, with responsibility for teaching Meteorology 205 and 216, Field Course and Meteorological Laboratory, respectively.[3] He did not immediately accept the offer. As he said,

> you see, Paula had to stay behind in Faribault when I came to Chicago. I couldn't support her on my meager salary. . . . So I was thinking of quitting the program. So I told Byers I would like to accept the position, but that I missed my wife. I asked him if he could give Paula a job. He said, 'We'll see what we can do.' He actually gave her a job. She was a map plotter. And she got to be in charge of all the map plotters. So she could plot a map and learn all the symbols better than I do. It was amazing. Of course, I then accepted the job. (Suomi et al. 1994)

---

3. Although Suomi is not listed in Figure 3.2, Byers said, "the Instrument Lab was successively under Mike Ference, Verner Suomi, and Earl Barrett" (H. Byers 1992, personal communication). Suomi is listed as an instructor during the 1943/44 academic year in Figure 3.1.

CHAPTER FOUR

Suomi's Research Style at the University of Chicago

As meteorology became central to aircraft operations in WWII, Rossby realized the great advantage that would come from a mobile upper-air unit. Midlatitude wind laws (geostrophic and thermal wind laws) in combination with upper-air observations of pressure or geopotential height would allow meteorologists to analyze atmospheric structure based on information at a single station—labeled "single-station analysis" by Rossby and described in technical reports by the University of Chicago meteorology staff members (Ference and Snodgrass 1942; Boyden et al. 1942; Jones et al. 1943).

The thermodynamic variables were obtained from the radiosondes and winds were obtained by tracking variously colored balloons—color dependent on sky conditions—with the aid of a sextant. Winds obtained in this manner were labeled "pibals" (an abbreviation for pilot balloons) that had come into use during the nineteenth century as guides for balloonists involved in racing. A group of cadets receiving instruction in pibal wind analysis is shown in Figure 4.1.

A Leeds and Northrup receiver collected the radio-transmitted data. The receiver was cumbersome and heavy, measuring 2 ft × 2 ft × 6 ft, and a truck was required to move the receiver from one location to another (Figure 4.2). Suomi had a mild interest in the receiver and decided to visit Rossby and talk about it. Frances (Fran) Ashley was Rossby's secretary and always "ran

Figure 4.1. Pibal (pilot balloon) class at Jackson Park for students in the University of Chicago's Cadet Program. Civilian instructors are shown in the foreground (Vincent Oliver on the left and Mildred Boyden), and one of the Air Force instructors is Joshua Holland (next to Oliver).

interference" for him. When Vern entered the foyer of Rossby's office and asked Ashley if he could have a few words with him, she said he was very busy and did not want to be disturbed. But Vern, in his polite yet forceful way, told Fran that it was important and could not wait. She reluctantly gave in to Suomi's request and led him to the door of Rossby's office, where Suomi remembered the event as follows:

> Well, Carl Rossby was not involved in science, the entire floor of his office was covered with checks and statements. He was trying to figure out his monthly bills! Anyway, he graciously let me talk about the mobile ballooning business. (Suomi and Lewis 1990)

Buoyed by Rossby's enthusiastic encouragement, Suomi marched to Michael Ference's office and discussed his ideas further. Within a few days, he informed Ference that he could reconstruct the receiver "for a song and fit it into a suitcase." Ference advised Suomi to seek funding from the U.S.

Figure 4.2. Truck used to transport the mobile radiosonde.

Signal Corps for the "suitcase receiver." Suomi's budget for the project was modest, and when the proposal was reviewed by the U.S. Army colonel in charge of meteorological research, he said, "I like the proposal, but up the budget by a factor of 10!" Vern had his first funded scientific project (Suomi and Lewis 1990).

The funded project included support for three meteorologists and two support staff, and research was to be conducted in the basement laboratory of Ryerson Hall—across the street from the meteorology building at 5727 University Avenue (Figure 4.3).

The receiver was to be tested at Jackson Park, just south of the campus along Lake Michigan's shoreline. Suomi's fellow graduate students Harry Moses and Earl Barrett were drawn into the project—both successful research meteorologists involved in instrument design following the war. This strong coupling between graduate students at the University of Chicago was the rule—"senior" students helping "junior" students with encouragement and instruction. The assistants were Charlotte Benton and [?] Flamenco. Among themselves and others in the Cadet Program, the group became known as the R3CT club. Harry Moses came up with the interesting formula. At ten o'clock every morning, Harry Moses would yell out, "it's coffee time" and the group would head to Hutchinson Commons for their midmorning break.

Figure 4.3. (Top panel) Building at 5727 University Avenue that housed the Meteorology Department at University of Chicago with Quadrangle Club's tennis courts in the foreground, and (Lower panel) science buildings along the University of Chicago's Quadrangle: Eckhart, Ryerson, and Kent Halls (front to back on right-hand side of the photograph).

He shortened his call to R3CT (Radiosonde Receiver Research Coffee Time). Since the proposal or the report of accomplishments cannot be located, we lack Vern's design for the receiver. But from Vern's oral history, we know that he and his team accomplished the contract goals (date uncertain). In his recollection he said, "I got it to fit in two suitcases instead of one" (Suomi and Lewis 1990).

In 1946, toward the end of the Radiosonde Receiver Research project, Vern became interested in measuring turbulence near the ground. He began to discuss this interest with Professor Starr and Oliver Wulf (both shown in Figure 3.6). These two scientists were exceptional for their clear thinking on theoretical issues. Wulf made his name at Caltech through investigation of ozone photochemistry and distribution of ozone in the atmosphere. He also had a strong interest in atmospheric optics (R. Bryson 1990, personal communication). At the recommendation of Rossby in 1941, Wulf was assigned to the Institute of Meteorology at the University of Chicago, where he taught Physics of the High Atmosphere to cadets (see Figure 3.2). Robert Millikan, executive director of Caltech, encouraged his professors to work outside academia—science in service to the world community (Goodstein 1991).[1]

Suomi, always the astute listener, overheard Wulf talking about acoustic measurements in the atmosphere. Wulf suggested it might be possible to measure the wind velocity by transmitting sound waves through the atmosphere. Wulf's argument rested on the assumption that sound speed would be increased when traveling with the wind and decreased when traveling against the wind. Suomi's discussion of this with Ference led to an unpublished survey article on sonic anemometry (Ference and Suomi, n. d.). Earl Barrett joined the project and a research proposal was submitted to the U.S. Army Chemical Weapons Division. Funding for the project was secured in 1946.[2]

Suomi and Barrett arranged access to a quiet, airtight room on campus and designed an experiment that would simulate the measurement of sound pulses in the presence of wind—a sonic anemometer. In a clever way of

1. An elaboration on Wulf's career can be found at http://www.nasonline.org/publications/biographical-memoirs/memoir-pdfs/wulf-oliver.pdf, a biographical tribute to Wulf by Harold Johnston (Johnston 2001).

2. The assumption that funding was received for this project comes from statements in Suomi (1957). Further, based on the oral history in Suomi and Lewis (1990), it is likely that the Ference and Suomi (n. d.) document was written in 1946.

thinking about the problem, Vern and Earl set up the sound source (small spark discharges) in the center of the room, and then placed receivers at different distances from the source—in effect, simulating the difference in speed due to wind. The experiment indicated that wind speeds as low as ~ 3 ′ 10^{-2} cm s^{-1} would be detectable by this process. As Vern said, "we had no high frequency microphones, the sound was generated by spark discharges, recorders based on 35 mm film pulled through an aperture at rates of 3–6 inches per minute, and then wrapping this film on a band saw blade that could spin and produce an audio frequency . . . by today's [1994] standards, we were working with archaic equipment" (Suomi et al. 1994). The technical report describing development of the sonic anemometer at the University of Chicago has not been located, but results related to the temperature measurement by the sonic anemometer appeared in Barrett and Suomi (1949). In 1953, the sonic anemometer was used in the O'Neill experiment that took place over the flat plains of Nebraska (Lettau and Davidson 1957). This instrument was used to measure turbulence near the ground, and it appears to be the first operational sonic anemometer—that is, a sonic anemometer used for an extended period of time in a field exercise. It is described in Suomi (1957).

CHAPTER FIVE

Professorship at UW–Madison: Early Years (1948–1953)

In 1947, Vern's role in development of the sonic anemometer landed him an invitation to visit Iowa State Agricultural College in Ames, Iowa [hereafter referred to as Iowa State University (ISU)[1]]. Faculty in agricultural meteorology at ISU wanted to know if Suomi's sonic anemometer would help give reasonable estimates of turbulent heat and moisture fluxes over their corn crop canopies (Figure 5.1). He worked with V. P. Starr several weeks before his visit to learn about the intricacies of turbulence theory. As Vern said, "we got down to fundamentals" (Suomi and Lewis 1990). The visit was so successful that Suomi was offered a faculty position at ISU. Almost simultaneously, Vern received a similar offer from UW–Madison. The offer from UW came in a letter from Reid Bryson, an assistant professor at UW–Madison who knew Vern when they were students in the Cadet Program at the University of Chicago. The positions and salaries from the two institutions were nearly identical—assistant professorships with salaries of $5500 per year at ISU and $5000 per year at UW–Madison. Where to go? With some indecision, Vern approached his close friend W. Ferguson Hall—a native of Iowa and the field research coordinator for the Thunderstorm Project at the University of Chicago (Figure 5.2). Vern asked, "Fergie, when people in Iowa go on vacation, where do

1. Iowa State Agricultural College was renamed Iowa State University in 1959.

Figure 5.1. (left to right) Robert Shaw, Victor Starr (kneeling), H. C. S. Thom, Verner Suomi, and W. H. Pierre shown examining instrumentation in a cornfield at the Iowa State University experiment station (ca. 1947) [courtesy of Iowa State College (University) Experiment Station].

they usually go?" "Wisconsin" was Hall's answer. Decision made! (Suomi and Lewis 1990). Vern was appointed Assistant Professor of Meteorology at UW–Madison in 1948 with affiliations in the College of Arts and Sciences and the College of Agriculture. Bryson's recollection of hiring Vern follows:

> In the spring of 1948 I had been talking to the Dean of Letters and Science [Mark Ingraham][2] for some time about bringing Verner Suomi to the University. There was considerable support from the College of Agriculture, especially after he had visited and demonstrated the kind of instrumentation he was developing, and its application to agricultural meteorology. When in mid-spring I suggested to the dean that in order for atmospheric science to develop there should be a separate department of meteorology at Wisconsin, Dean Ingraham said, in essence, "Okay, you are it effective July first." There had been reluctance on the part of the Geography Department's chairman to

2. Ingraham had been involved with the Cadet Program through service as chair of one of the committees (Undergraduate Training Committee) (Rossby 1943).

Figure 5.2. Thunderstorm Project personnel (left to right): Ferguson Hall (Field Research Coordinator), Colonel Lewis Meng (Air Force Operations), and Horace Byers (Project Director). Background: Northrop P-61 (Black Widow), a radar-equipped WWII fighter plane used for flights through thunderstorms (ca. 1947) (courtesy of Douglas Allen).

> add more staff in meteorology. I then asked whether I could recruit Suomi. I was pleasantly surprised when he said 'yes.' . . . On the 15th of July I drove to Chicago to bring the Suomi family to Wisconsin.
>
> —R. Bryson (Suomi Memorial, 1995, pp. 9–10)

With only two faculty members in the newly founded department, the course load was heavy. We do not have reminiscences from students in his classes during the 1940s, but the following statements from his colleagues and students from later decades attest to his teaching style.

Frederick House (2014, personal communication):

> In [my] five years of graduate school (1960–1965) [at UW–Madison], Professor Suomi never taught a course taken by graduate students. Every morning at about 10:00 AM, his graduate students would gather outside his office and we would go to the Student Union across the street for coffee. This was his time for teaching graduate students. The many ideas coming out of his head, non-stop, were remarkable.

Steve Cox (2015, personal communication):

> Graduate level classes that Verner taught [in the late 1960s] typically featured a contemporary research problem that students collectively pursued over the course of the term. Many of these class projects resulted in papers published in the refereed literature.

Professors Donald Johnson, William Smith, and John Young (Suomi Memorial, 1995, pp. 6–7):

> He [Suomi] loved classroom teaching of undergraduates, and many considered his classes unforgettable. He asked students for curiosity, common sense, and positive attitudes. In return they got spirited explanations of complex phenomena and simple ideas for applications. Ultimately, his ways of thinking took precedence over detailed content. His teaching was no product of fixed procedures; it was an unrepeatable process that was a window into a mind in constant motion. By his example, students learned to inquire more boldly and effectively.

Martin Hoerling (2010, personal communication):

> My own recollection of the man—from the perspective of a student in the late-1970s, who took Professor Suomi's satellite meteorology class—was of a person akin to a spark plug, firing on all cylinders all the time with his thoughts/concepts/ideas. For those of us that just wanted to learn the basics, you might imagine a bit of youthful academic frustration—we wanted a mere synthesis (that could be repeated on tests for the sake of a decent grade!)—yet, Professor Suomi delivered innovation and forward looking perspectives and then delivered tests that seemed to be sketches of his prior evening's thoughts drafted atop a coffee-stained diner napkin!

In addition to his teaching responsibilities, Suomi made connections across academic lines that included links to limnology, soils, agriculture, and electrical engineering. He was especially interested in linking with electrical engineering where kindred spirits resided—most notably, Professor Robert Parent, a recognized expert in electronic instrumentation systems and data processing equipment.

Suomi never lost interest in agricultural meteorology after his visit to ISU in 1947 (Figure 5.1). And with his joint academic appointment that linked him to the College of Agriculture, he began an experiment at the Marsh Farms on the far western edge of Madison—an experiment designed to calculate the heat budget over a cornfield. Motivation for the study came in part from development of an accurate net radiation sensor (Suomi et al. 1954). The basic tenet of the experiment was that net radiation would balance the surface sensible and latent heat fluxes and heat conduction into the ground during the summer months when the corn was growing. The experiment was conducted over four days in 1952 (1–4 September). The accuracy of a net radiometer was central to the success of the experiment, as stated in the publication that described the experiment (Suomi 1953, p. 9):[3] "The net radiation term [in the budget equation] measured with this instrument was accurate to about two percent. This is indeed fortunate, since radiation is usually the largest term of the heat budget."

The 49-page document describing the experiment (Suomi 1953) appeared to be written hastily, where more details and elaboration at key junctures would certainly have been appreciated by the typical reader. Nevertheless, the description of the equipment and accuracy of measurement was well presented. Since the latent heat flux is a major term in the budget equation, Suomi creatively designed an evaporimeter to measure the moisture transport from the cornfield to the atmosphere. A standard evaporimeter is a large and deep cylindrical tank (66 inches in diameter in Suomi's case) and 3–4 feet tall where depth of water is measured by a micrometer or hook gauge affixed to the inside of the tank. The mass of water evaporated from the tank in a given period of time multiplied by the latent heat of vaporization for water delivers the latent heat flux from the tank. Suomi floated a smaller cylindrical tank (60 inches in diameter and 20 inches deep) within the evaporimeter and filled this smaller container with blocks of soil and embedded corn stalks from the field. The surface of the water in the annulus (moat) was covered with oil to prevent evaporation of water from the evaporimeter. Evaporation from the soil and stalks, however, reduced the weight of the inner container's contents. The mass of evaporated water could be estimated by recourse to Archimedes' principle and used to calculate the latent heat flux from the soil and cornstalks. The resolution was sufficient to get accurate measurements of water loss over a period of several hours.

---

3. This publication was submitted to the faculty of the Division of the Physical Sciences at the University of Chicago in June 1953 as part of his candidacy for the PhD degree. We have found no other report or publication related to this experiment.

Table 5.1. Average surface winds and temperatures at Truax Field (Madison, WI) for September 1–4, 1952.

Date	Temperature (°F)	Wind direction	Wind speed (mph)
1	60	WNW	8.8
2	56	WNW	14.3
3	59	WNW	9.6
4	64	SSW	9.9

When questioned about the experiment nearly forty years later, his first reaction was "I had a 40% error in the budget on the last day [September 4, 1952] . . . I finally figured it out. On that day there was strong warm air advection and the model didn't account for horizontal processes . . . that warm air advection heated the soil" (Suomi and Lewis 1990). Table 5.1, constructed by the authors, indicates a wind shift to the south–southwest over Madison on September 4, with a marked increase in surface temperature at the airport in Madison—most likely due to warm air advection.

Dean Ingraham, the dean who was instrumental in establishing the UW–Madison meteorology department, as mentioned earlier, also played a key role in advancing Suomi's career. He visited Suomi soon after the experiment at Marsh Farms was completed. Ingraham said, "you've got to get your PhD" (Suomi and Lewis 1990). Vern had not seriously considered getting the PhD, although he had completed the required coursework at the University of Chicago in the 1940s (Suomi et al. 1994). But in response to the dean's admonition, Suomi sent the 49-page report to the University of Chicago's meteorology department. He hoped it would find favor as partial fulfillment of PhD requirements. And the following policy at the University of Chicago fit Vern's situation perfectly. The policy read:

> Select a thesis topic and begin the investigation for the dissertation under the supervision of a faculty member. The student, as part of the demonstration of originality, will be expected to propose his own thesis topic for acceptance by Chairman of the Department, and will also be responsible for finding a sponsor for his research. (Announcements, University of Chicago, 1950)

Norman Phillips, who took the B.S. (1947) and Ph. D. (1951) in meteorology at the University of Chicago, had the following opinion on this requirement: "This policy had Rossby's fingerprints all over it" (N. Phillips 1991, personal communication).

Rossby was Suomi's advisor before he left Chicago to become director of the Swedish Meteorological and Hydrological Institute (SMHI) in 1947.[4] Rossby continued to commute between SMHI and University of Chicago for several years but left supervision of Suomi's dissertation to Horace Byers, chairman of the Department of Meteorology, after Rossby's departure. Byers was impressed with Suomi's study of the heat budget over a cornfield and scheduled the dissertation defense for June 1953.

---

4. Following WWII, Rossby wanted to attract the best meteorologists in the world to his institute, including those from Axis countries (Germany, Italy, and Japan) and Russia. When the University of Chicago exhibited unwillingness to support such an institute, he left Chicago and became the founding director of SMHI (A. Wiin-Nielsen 1993, personal communication).

CHAPTER SIX

Epiphany at Chicago

Suomi passed the doctoral exam at Chicago without a problem, but he never forgot one of the comments and one of the questions that followed his presentation. The comment was made by Professor Sverre Petterssen as a reaction to Vern's explanation of the 40% error in the heat budget on September 4. During the question period following the examination, Vern explained the error as follows: "it was the soil advection term that led to the discrepancy" (Suomi and Lewis 1990)—a wording blunder that typifies the response of a doctoral student under interrogation. Petterssen said, "The only way you can advect soil is with a wheel barrow." The serious question came from Herbert Riehl (Figure 6.1). The question: "So now you've examined the energy budget over a cornfield, how would you go about examining the heat budget for the Earth and its atmosphere?" (Suomi and Lewis 1990).

One can only imagine the "fireworks" that went off in Suomi's head after he heard and considered this question. However, based on his memory of the incident and the excitement in his voice when he recounted the event during an interview (Suomi and Lewis 1990), one can presume that he had an epiphany—a sudden flash of recognition that became deep-seated with time. He immediately recognized that the calculation of the budget over a cornfield could now be applied to the globe. And the observations would

Figure 6.1. Herbert Riehl (1915–1997) in a contemplative mood while sitting atop Montgomery Peak, White Mountains, on the California–Nevada border (June 1961) (courtesy of H. Klieforth).

not be ground-based, but made from space.[1] With his knowledge of net radiation in theory and practice, Vern realized that radiative processes would dominate other processes in the budget equation for the Earth–atmosphere system. In that sense, the calculations of heat budget over the globe would be theoretically simpler than the heat budget over a cornfield. However, Suomi now needed to design a net-flux radiometer that could make measurements while affixed to a rotating platform in space.

The following paragraph is excerpted from Lewis et al. (2010):

> With Riehl's question continuing to ring in Suomi's ears and with the exciting cloud pictures that were being transmitted from camera-equipped high-altitude rockets such as the German V-2 and U.S. Navy's Viking, Suomi began to think about sensors that could be placed on spacecraft. Further, he was impressed with Henry Houghton's comprehensive study of the heat budget of the Northern Hemisphere (Houghton 1954). A prescient thought

1. It is probable that Riehl mentioned satellite observations in discussion with Suomi during the PhD defense. Riehl was thinking about satellite observations in support of his work on general circulation between the equator and poles.

> arose in Suomi's mind: Instead of reliance on ground-based data for such a study, why not measure the Earth–atmosphere radiation budget from space? His enthusiasm soared when he heard Joseph Kaplan give a Sigma Xi lecture at UW–Madison in 1956. Kaplan, UCLA professor of physics who was instrumental in bringing meteorology to UCLA in 1940, was the national chairman of the International Geophysical Year (IGY) committee (Day 1999). In his talk, he presented plans for the IGY, which was scheduled to take place between 1 July 1957 and 31 December 1958. Suomi told Kaplan he had been working on a sensor to measure the Earth–atmosphere radiation budget from space. Kaplan encouraged Suomi to contact Harry Wexler . . . the national chair of the IGY meteorology committee. And although the deadline for submission of proposals to the IGY had passed by several weeks, Wexler "squeezed" Suomi's proposal into the mix. He also added, "Keep it simple Suomi" (Suomi, personal communication, 1984)[2]. Wexler worked with Suomi on the proposal and sought advice from Professor Riehl.

The letter exchanges between Wexler and Riehl are found in Figures 6.2 and 6.3.[3] Riehl's letter is wide ranging but with definite concentration on the value of satellite radiation measurements in support of understanding the physical processes that drive general circulation. From Riehl's letter:

> The net radiation over the whole belt [30S – 30N]. . . will be highly important in deciding whether, by and large, longer-period fluctuations of middle latitudes circulations arise, or are related to fluctuations of the net radiation, or whether they are due to internal mechanisms.

In essence, Riehl's statement is at the heart of research that came with the Earth Radiation Budget Experiment (ERBE), as discussed in chapter 9. He also hits near the target center when he muses about the annual variability of the general circulation, again discussed in chapter 9. These prescient statements served as a guide for Suomi beyond the IGY. The strength of Riehl's arguments paid dividends when Wexler presented Suomi's case to the IGY Satellite Panel on Monday, December 3, 1956. On January 16, 1957, the IGY

---

2. Throughout his career, Vern would often use this phrase in discussions with his students and colleagues—he called it his KISS principle.

3. The letters in these figures are not the original letters. The original letters were retyped to make them more legible.

THE UNIVERSITY OF CHICAGO
Chicago 37, Illinois
Department of Meteorology

Nov. 28, 1956
(Somewhere over the Rockies)
(Nati Ann running interference)
(Plane has mechanical shakes, hope you have bi-focals)

Dear Harry:

Some hours have gone since our early morning encounter, but they have been enough for my latent astonishment to your remarks over satellites to solidify. My knowledge comes only from magazines, etc. Various articles have led me to think that the matter of radiation calculations via satellite was an accepted part of the program, fully worked up and integrated. Apparently, this is an impression gained from excessive reporting, as so often (but how can one judge?), and the subject as yet is rather far from passed on. I'll presume this to be the situation.

In my estimate, net radiation measurements from a satellite oscillating between (what is it?) 30°N and S would be of decisive importance for some aspects of our science -- if your justification must be practical, I suggest quoting long-range forecasting. Of course, there should be independent measurements of long and short wave radiation, and of the Spectrum (energy) of the short wave radiation. Further it would be desirable to have values transmitted, if possible, along the path of the moonlight. However, the net radiation over the whole belt, if that is all that one can determine, will be highly important in deciding whether, by and large, longer-period fluctuations of middle latitude circulations arise, or are related to, fluctuations of the net radiation, or whether they are due to internal mechanisms within the atmosphere.

Taking the tropics as the source region of heat, we can calculate at least, using satellite data:

(1) The net heat received
(2) The flow through latitudes 15°N and 30°N

This leaves open:
(3) The same flow on the southern hemisphere unless IGY data are enough down there
(4) Storage in tropical oceans, probably not a major factor except seasonally (we hope).

As I said this morning (plane shakes more, now atmospheric turbulence) 1954 CST, near Grand Junction, Colo.), the heat flow through lat. 15°N and 30°N varies strongly in one winter, from nothing to 200% of climatic flow needed to balance heat loss of higher latitudes. Period: order of 2 months. Fluctuations are so large that really hardly in doubt. Additional evidence available. At least, one can correlate the heat export with satellite data, and see whether net radiation has swings on the times scale of the heat export, and at the right time. Complete heat budget, or course, would be still better.

I think this is fundamental information for guiding meteorological research on long (and very long) period changes, also methods for computation schemes. Suggest three programs:

(1) Satellite radiation measurements (as discussed).
(2) Heat flow through lat. 15° and 30°N (and S if possible)
(3) Available parameters of middle latitude flow to be correlated, means around globe and regional

Well, I hope these comments are of some use, and you have some eye sight left. Get secretary to type for you. I think, this is very important. If goes through, would like to apply for participation on analysis.

Happy days with the penguins,

Sincerely,

/a/ Herbert

Figure 6.2. Letter from Herbert Riehl to Harry Wexler (originally handwritten but later typed) related to issues in Suomi's research plan for the IGY (courtesy of Schwerdtfeger Library, SSEC, UW–Madison).

December 12, 1956

R-3

Dr. Herbert M. Riehl
The University of Chicago
Department of Meteorology
Chicago 37, Illinois

Dear Herbie:

Many thanks for your letter written under rather trying circumstances on your flight west on November 28.

Your support of this project was quite welcomed and I was able to use some of your arguments in presentation of our case before the IGY Satellite Panel on Monday, December 3, together with Vern Suomi.

Your project has now been accepted by the IGY and should be on the vehicle scheduled to be placed into orbit sometime in the spring of 1958.

Vern Suomi is beginning to work furiously on establishing a realistic prototype by July 1st. By copy of this letter I am asking Vern to send you whatever available information he has on the proposed instrumentation to be installed on the satellite. I am also transmitting to you with this letter a reprint and also a copy of the original proposal I made last June 7 when I first appeared before the IGY Satellite Panel pleading for specs on a vehicle.

Hope you enjoyed your double Christmas!

Sincerely yours,

Harry Wexler
Director of Meteorological Research

cc: Prof. V. Suomi
Dr. S. Fritz

Figure 6.3. Letter from Wexler to Riehl in response to letter shown in Figure 6.2 (courtesy of Schwerdtfeger Library, SSEC, UW–Madison).

national committee accepted Suomi's proposal.[4]

My father, a mathematician, once shared this offhand remark, at a time when I was still in school. He certainly wasn't claiming this to be an original thought; it was simply an observation he made during whatever conversation we were having at the time. Ever since, I've reflected on the wisdom of this. Early on, for example, when I was finishing up my graduate work at the University of Chicago, I realized that in my thesis I'd stumbled on a small problem in an obscure corner of atmospheric science that had never been tackled before . . . and yet was linear. The mathematics was a breeze.

4. An elaboration on the role played by Wexler in post-WWII meteorology is found in Fleming (2016).

II

Earth's Heat Budget from Space

CHAPTER SEVEN

Suomi–Parent Ping-Pong Radiometer and Its Principle of Operation

As cited in the last paragraph of Wexler's letter, Vern had already begun to develop a prototype space-borne radiation sensor—humorously called a "ping-pong radiometer" because the small plastic balls used in table tennis resembled the radiometer's spherical sensors. Suomi had drawn his close colleague into the project—Robert (Bob) Parent, UW–Madison professor of electrical engineering. By all accounts and reminiscences, these two men were well matched, and there was a complementarity in their approach to problem solving—Vern the idea man and Bob the gifted practitioner. There are several stories about Suomi and his idea generation. Joost Businger said it this way: "Suomi was a scientist with many ideas . . . but Vern's ideas needed to be filtered and the good ones selected" (Businger 2005). James Weinman, who worked for Vern as a postdoc in 1963, put a quantitative measure on the ideas: "I was given the task to trouble-shoot Suomi's ideas. 80% of Suomi's ideas were worthless, but the other 20% were damn good" (Weinman 2003). In the particular instance of Parent working with Suomi, Paul Menzel (Figure 7.1) viewed the collaboration as follows (2013, personal communication):

> They made things happen when they worked together—creative minds combined with strong personalities. Vern was coming up with many ideas but Bob was working on the last idea that had taken hold. Bob was building

Figure 7.1. Paul Menzel, 1990 (courtesy of SSEC, UW–Madison).

> to that design. Then Vern comes in and says 'Drop everything, Bob, I've got a new idea how to do this.'

This was further supported by Tom Vonder Haar (2014, personal communication):

> On the subject of instruments and Bob Parent, I remember interacting with Bob once I became a post-doc. He was long suffering. Trying to keep Vern from tearing down the thing that was working. Suomi always had a better idea, but Bob was trying to keep it on track and within budget and within performance. He worried about Suomi getting it too complicated or getting rid of things that already worked and getting a new gadget.

Figure 7.2. Suomi and Parent at their workbench (courtesy of UW–Madison Communications).

Despite this aspect of Suomi's nature, Parent knew that in combination they were greater than the sum of their independent parts—an "inequality principle" of sorts. These two men worked tirelessly on the prototype, where many nights were spent at the lab with sleeping bags for occasional rest, a trip to one or the other's house for morning breakfast, and then interleaving classroom teaching and advising with continued work on the radiometers (this style of collaboration would continue throughout their careers)

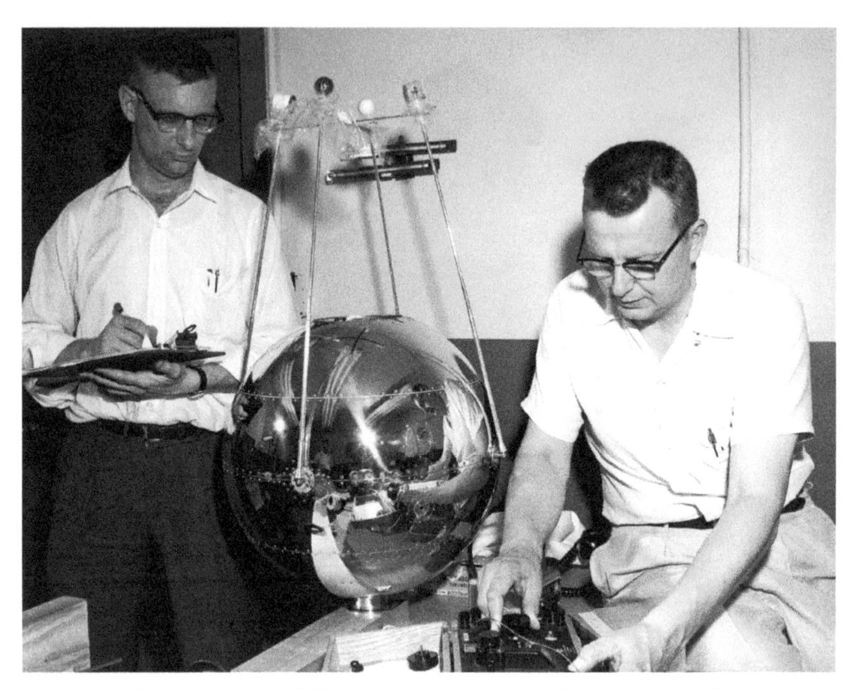

Figure 7.3. Charles R. Stearns (left) and Verner Suomi with the Vanguard satellite that carried Suomi's early heat balance experiment in 1959. This launch was unsuccessful. Circa June 1959 (courtesy UW–Madison Archives).

(P. Calloway 2009, personal communication).[1] In Figure 7.2, they are shown at the workbench, and in Figure 7.3, student Chuck Stearns is seen working with Suomi. In view of Parent's extraordinary contributions (Figure 7.4), we have included an expanded summary of his career in the Appendix (Robert J. Parent's Vita).

Vern's retrospective thoughts on the radiometer are found in the transcript of his oral history:

> The black sphere would see the solar, terrestrial, and reflected radiation while the white sphere would see terrestrial radiation. It would be better if you had a flat plate normal to the radiation but that would be impossible on a satellite that's spinning. Now there are three places on the earth that are different. One of them is on the dark side of the earth where you only

---

1. Patti Calloway is the daughter of Robert Parent, and this memory comes from her experience as a young woman talking to and observing her father and Suomi at breakfast.

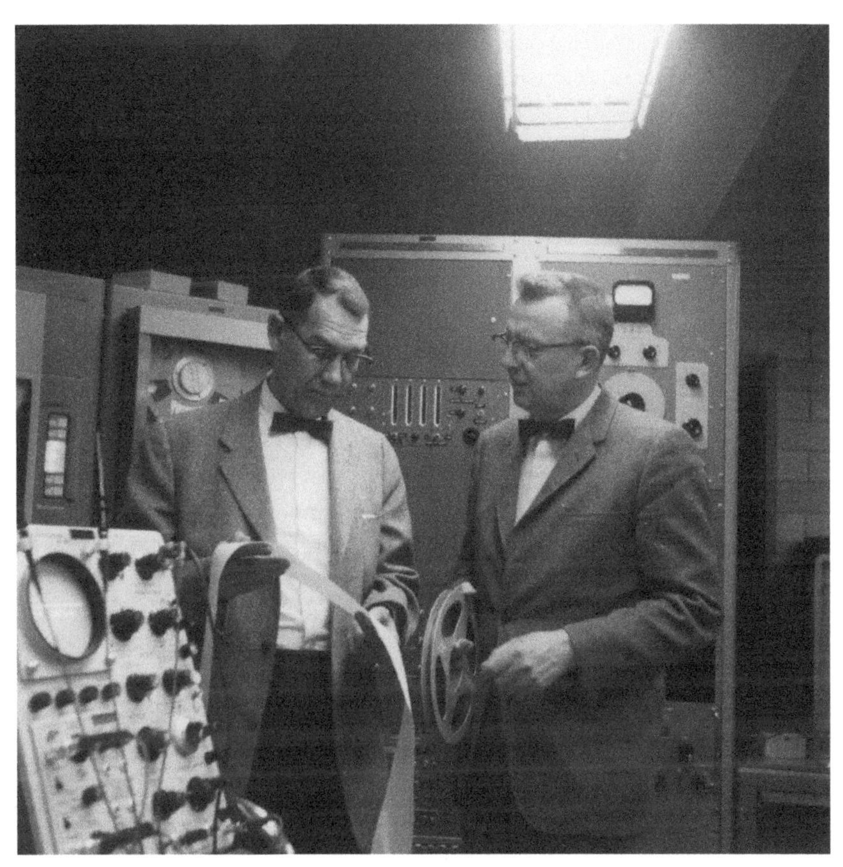

Figure 7.4. Robert J. Parent (left) and Verner E. Suomi reading data tape, circa 1959 (courtesy of UW–Madison Archives).

> have longwave [terrestrial] radiation, on the sunny side of the earth, you have longwave radiation, you have reflective radiation, and you have the sun shining on the device. And then there is the part of the orbit which is beyond the sunset point, which is in darkness as far as the sun is concerned, but the sensor, or satellite is still up in the sunlit portion of the orbit [above the atmosphere]. So you have the sun and earth radiation by itself [the flux of reflected solar radiation is absent] . . . and heavens, that's three unknowns, and three equations and there was nothing to solve it. (Suomi et al. 1994)

Budget Equations for the Radiometers

As typical of Suomi, his explanations are appealing but a little cavalier. To clarify his line of argument in the quote above, we introduce the budget

Table 7.1. Properties of the ping-pong radiometers. The "W" refers to low absorptivity/high emissivity response to the radiation (both visible and infrared), and "B" refers to high absorptivity/high emissivity response to the radiation.

Coating	Response to visible: α	Response to thermal infrared: ε
Black	B ($\alpha_B = 0.90$)	B ($\varepsilon_B = 0.90$)
White	W ($\alpha_W = 0.10$)	B ($\varepsilon_W = 0.90$)

equations associated with this problem.[2] Let us first define the properties of the sensors, where "W" refers to surfaces with low absorptivity and high emissivity and "B" refers to surfaces with high absorptivity and high emissivity. Entries in Table 7.1 give the radiative properties of the two color-coated spheres. Absorptivity and emissivity are represented by α and ε, respectively, and numerical values of these parameters for the two spheres are included in the table. Subscripts B and W denote properties of the black-coated and white-coated spheres, respectively.

Further, we define the fluxes Φ (W m^{-2}) and their associated spectral ranges as follows:

- Φ_S, flux of solar radiation (0.2–3 μm);
- Φ_{IR}, flux of thermal infrared radiation (3–50 μm); and
- Φ_{rS}, flux of reflected solar radiation (0.3–3 μm).

The radiative heat transfer equation for each sphere is given by

$$H\frac{\partial T}{\partial t} = \alpha \cdot (\Phi_S + \Phi_{rS}) + \varepsilon \Phi_{IR} - 4\varepsilon \cdot \sigma T^4$$

where H is the thermal capacity of the sphere per projected area (J m^{-2} K^{-1}), σ is the Stefan–Boltzmann constant (elaborated on in chapter 9), and T is the temperature of the sphere (K). The factor of 4 stems from the fact that the surface area of the entire sphere ($4\pi r^2$) emits infrared radiation but only the projected area (πr^2) absorbs radiation. In other words, the equation describes time rate of change of sphere temperature in response to heat input (solar, reflected solar, and infrared flux of radiation) and heat output (flux of Stefan–Boltzmann radiation). As already mentioned, subscripts B and W

2. Further information on Suomi's view of the problem is found in http://library.ssec.wisc.edu/SuomiWebsite/SuomiImages/scanned%20documents/Suomi_IGY_1957_032.pdf.

are attached to the variables H, T, α, and ε to distinguish values associated with the colored coatings.

There are two equations—one for the white-coated sphere and one for the black-coated sphere—for each satellite orientation: lighted side, dark side, and the side "beyond the sunset point" (i.e., a total of six equations). The time rate of change of the temperature of each sphere is measured with a thermocouple (known quantities). Consequently, the unknowns in the heat transfer problem are the three fluxes. In line with Suomi's explanation above, one could take the measurements for the white sphere at the three orientation positions mentioned, and this would yield three equations in the three unknowns (the fluxes). But to overcome the random error in measurements, it makes sense to include more than three equations in the three unknowns. This problem with more equations than unknowns is solved by the method of least squares. It is solved as a least squares problem where the governing equations are not required to be satisfied exactly, but in such a way that the departure from exactness is minimized (see Lewis et al. 2007).

A discussion that complements this review of radiation budget measurements, and the Wisconsin instrument development in particular, is found in chapter 4 of *Remote Sounding of Atmospheres* (Houghton et al. 1984).

CHAPTER EIGHT

Explorer VII: The Magnificent Voyage

The first two rockets designed to place a satellite in orbit with attached ping-pong radiometers failed. In February 1959, the Vanguard rocket carrying the Vanguard Radiation Balance satellite crashed into the Atlantic Ocean soon after takeoff from Cape Canaveral. Six months later in July 1959, a Juno II rocket designed to place the radiometer-equipped *Explorer VI* satellite into orbit was destroyed immediately after liftoff at the Cape. The 76-foot rocket bolted upward for just a few seconds before falling to Earth in a ball of orange flames. The failure occurred in the usually reliable first stage of the rocket. Suomi and Parent were in the blockhouse that was 50 meters from the launch pad. The front-page headlines in the Wisconsin State Journal (July 17 edition) read "Suomi Picks Up Pieces, Hopes for a Third Try." Indeed, after the fire was extinguished and the wreckage cooled, Suomi and Parent hacksawed their instrument package from the nose cone. In this same article, Juno II project manager Herman LaGow (shown in Figure 8.1) was quoted as saying "the satellite launching attempt may be rescheduled in a few months."

A third attempt to insert the satellite into orbit was scheduled for October 1959, a little more than two months after failure of the Jupiter Intermediate-Range Ballistic Missile (IRBM). Soon after the second failure and in order to prepare for the next launch, Suomi and Parent visited Werner von Braun at the Redstone Arsenal in Huntsville, Alabama. Von Braun had been proj-

Figure 8.1. (left) Suomi and NASA engineer Herman LaGow working on *Explorer VII* (courtesy of NASA, 1959).

ect director of the Ordnance Guided Missile Center at Redstone Arsenal since 1950 and had supreme power in decisions related to rocket and missile launches in the United States (Neufeld 2007). Here, Suomi and Parent faced a major hurdle. In essence, von Braun objected to having the radiometers placed at the ends of extended antenna booms. Based on analyses of the two failed launches, it would be hard to place any blame on radiometer placement.[1] Yet, von Braun's stubborn stand could well be related to the extreme pressure he and his team were facing at that time—pressure in large part due to uncertainty in the Redstone Arsenal's future (Neufeld 2007, chapter 14).[2] As recounted by Suomi,

> Bob Parent and I went [to the Redstone Arsenal in Huntsville, Alabama] to talk to [Werner] von Braun about putting the radiometer on board Explorer

---

1. However, the high rate satellite spin with radiometers attached to the ends of antennae could be a destabilization factor.

2. In 1960, von Braun's rocket development center at Redstone was transferred from the Army to the newly established civilian agency, the National Aeronautics and Space Administration (NASA). With the transfer, von Braun became director of NASA's Marshall Space Flight Center in Huntsville, AL, on the same grounds as the Redstone Arsenal.

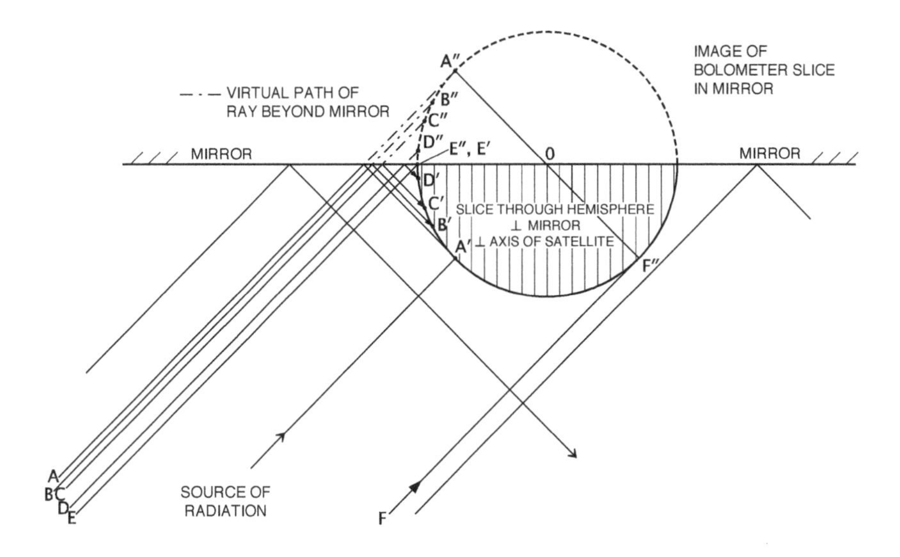

Figure 8.2. Schematic diagram showing the action of a radiation current intercepted by the Suomi–Parent hemispheric bolometer backed by a mirror.

> VII. He [von Braun] said, "no way [can you put the spheres on the end of the antennae]." I was done. We were dejected. Later that night in our hotel room, I noticed a bowl of oranges in the room. I cut one in half and put it against the mirror—a full orange! Next day I proposed to von Braun that we have mirrors on the underside of the hemispheres. The image of the hemisphere which appears in the mirror makes the sensor look like a full sphere. He accepted the idea. (Suomi et al. 1994)

Vern adapted the design by arguing that mirror-backed hemispheres, by virtue of the satellite's spin, would act similarly to isolated spheres in space insofar as radiation interception was concerned (Suomi 1961).[3] Figure 8.2 is a schematic diagram that captures the essence of Suomi's argument. In this figure, we assume the radiation source (shortwave or longwave radiation) emanates from the lower-left corner of the figure, and its distance from the radiometer renders the radiation current uniformly parallel. The rays *AA″*, *BB″*, *CC″*, and *DD″* impact the planar mirror and are reflected (where the angle of incidence equals angle of reflection) onto the hemisphere at *A′*, *B′*, *C′*, and *D′*, respectively—thus, the incidence of these rays on the hemisphere

3. This paper is available online at http://library.ssec.wisc.edu/SuomiWebsite/SuomiImages/scanned%20documents/Suomi_ExplorerVII_Radiometer_018.pdf.

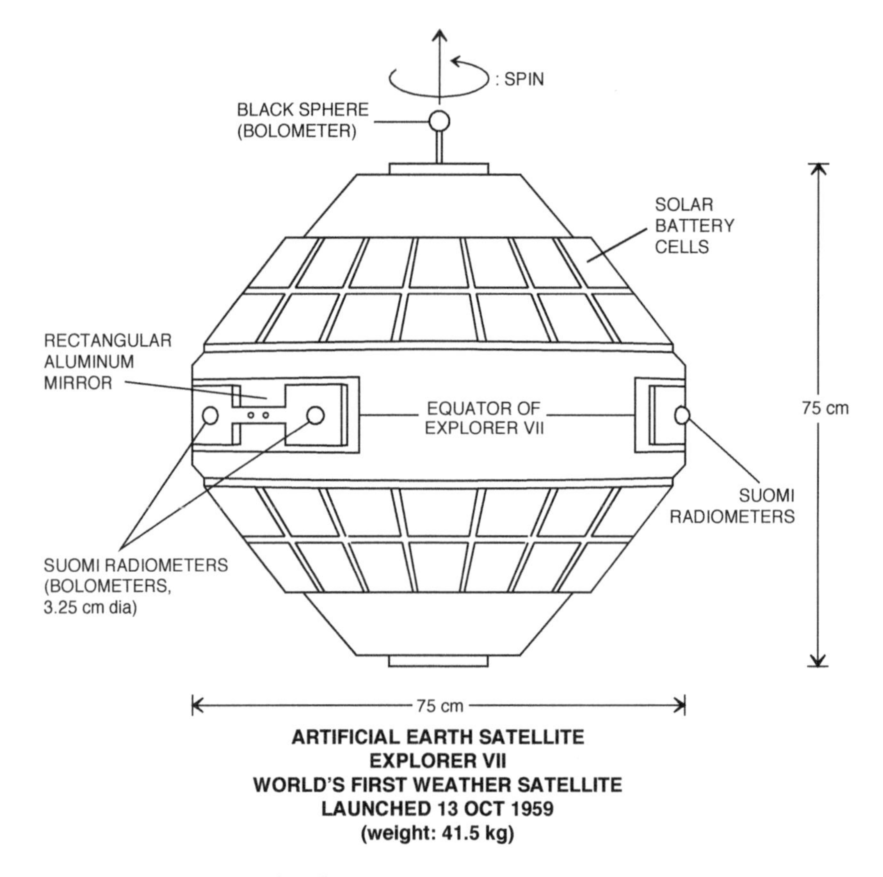

Figure 8.3. Components of *Explorer VII.*

leads to the same radiative input that would occur on a sphere in the same space as the hemisphere and its mirror image as long as the planar mirror is much larger than the hemisphere.

A schematic drawing of the *Explorer VII* satellite is found in Figure 8.3, where the hemispheric radiometers and underlying rectangular aluminum mirrors are positioned along the satellite's equator. A photo of two hemispheric radiometers (bolometers[4]) and the mounting surface (with mirrors) is shown in Figure 8.4. In addition to the black- and white-coated hemispheres, Suomi and Parent had the foresight to include gold- and "tabor"-coated hemispheres.[5] A black sphere was mounted on the axis of the satellite

4. A bolometer is an instrument that measures the intensity of radiation by employing a thermally sensitive electrical resistor (Glickman 2000).

5. See Tabor (1956) for the interesting research that led to the "tabor" coating.

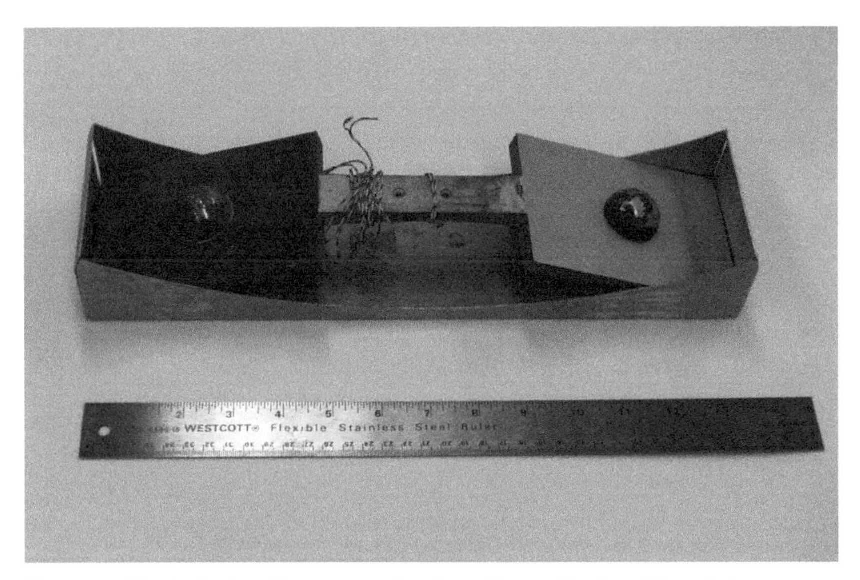

Figure 8.4. Hemispheric radiometers on flat plates. The positioning of these bolometers around *Explorer VII*'s equator is displayed in Figure 8.3 (courtesy of SSEC, UW–Madison).

at the top. It was used to detect any deterioration in the mirror surfaces by comparison with the blackened hemisphere. Thus, there were five radiometers on the satellite. The color coats complemented the radiative properties of the black- and white-coated hemispheres). As stated in the previous chapter, an increased number of radiative balance equations serve to improve the accuracy of the fluxes when the system is solved using least squares. Following the symbolism B and W for absorptivity and emissivity used in Table 7.1, the response of the four hemispheres and the single sphere to visible and thermal infrared radiation is displayed in Table 8.1.

The bolometers were thinly plated with silver and their temperatures were measured by glass-coated bead thermistors in solid contact with the hemispheres. The sensor temperatures were telemetered back to Earth. Since

Table 8.1. Properties of bolometric coatings where "B" and "W" have same meaning as in Table 7.1.

Coatings	Response to visible	Response to thermal infrared
Tabor	B	W
Gold	W	W
Black	B	B
White	W	B

there were no data storage recorders on board *Explorer VII*, a radio transmitter sent data to ground stations.[6] Interestingly, Soviet scientists were able to access data when the satellite passed over their ground stations (Malkevich et al. 1963).

The instrument system and sensors are described in detail in Suomi (1961). Discussion of data handling and telemetry gives an idea of the precision needed to make this instrument package work. The huge volume of data generated required automatic data handling wherever possible, and in a month's time, 432,000 separate measurements were transmitted.

6. There is interesting verbatim discussion by Suomi of the *Explorer VII* results at a NASA press conference in December 1959. This discussion can be found online at http://digital.lib.uiowa.edu/cdm/ref/collection/vanallen/id/2893.

CHAPTER NINE

Earth's Radiation Budget from Satellites: Theme of the 1960s

Review of Earth–Atmosphere Radiation Theory

Historically, research into atmospheric radiation is temporally uneven. In the late nineteenth to early twentieth century, empirical methods led the way in trying to understand the effect of CO_2 (carbon dioxide) and H_2O (water vapor) on so-called greenhouse warming. Science historian James Fleming has admirably told the story of these early efforts (Fleming 1998, 2007). The major contribution to radiation theory in gaseous atmospheres came from celebrated astrophysicist Subrahmanyan Chandrasekhar (Chandrasekhar 1950). The fundamental physical constraints that governed this radiation theory were the basis for the specialized transfer equations developed by Suomi and Parent for the ping-pong radiometers and the hemispheric bolometers (discussed in chapters 7 and 8). With the advent of numerical weather prediction (NWP) in the mid-twentieth century, coupling between radiation theory and atmospheric dynamics commenced (see chapter 15 for details).

Chandrasekhar's theory is also the basis for understanding the radiative equilibrium of the Earth–atmosphere system (also see Goody 1964; Goody and Walker 1972). Let us briefly present the salient features of this equilibrium theory as it applies to the Earth's heat balance. Essential components of the theory are displayed in Figure 9.1. One of these components, the Stefan–Boltzmann law, was developed over a period of nearly 70 years—first

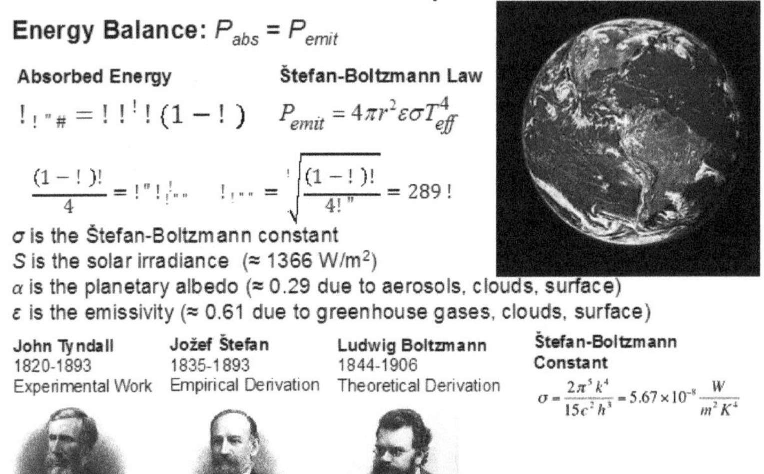

What Determines the Effective Temperature of the Earth?
Energy Balance: $P_{abs} = P_{emit}$
Absorbed Energy
Štefan-Boltzmann Law
$P_{emit} = 4\pi r^2 \varepsilon \sigma T_{eff}^4$
= 289
σ is the Štefan-Boltzmann constant
S is the solar irradiance (≈ 1366 W/m²)
α is the planetary albedo (≈ 0.29 due to aerosols, clouds, surface)
ε is the emissivity (≈ 0.61 due to greenhouse gases, clouds, surface)
John Tyndall
1820-1893
Experimental Work
Jožef Štefan
1835-1893
Empirical Derivation
Ludwig Boltzmann
1844-1906
Theoretical Derivation
Štefan-Boltzmann Constant
$\sigma = \frac{2\pi^5 k^4}{15c^2 h^3} = 5.67 \times 10^{-8} \frac{W}{m^2 K^4}$

Figure 9.1. What determines the effective temperature of the Earth? Photographs of Tyndall, Stefan, and Boltzmann are shown at the bottom of the figure.

by experimentalists Dulong and Petit (1817) and Tyndall (1868), followed by Jozef Stefan's (1879) empirical work with the experimental data, and finally the theoretical derivation by Stefan's doctoral student Ludwig Boltzmann (1884). Tyndall, Stefan, and Boltzmann are pictured at the bottom of Figure 9.1.

Suomi wrote that, "except for long-term climatic change, the energy the earth intercepts [i.e., absorbs in the nomenclature we have adopted] from the sun over a period of years is balanced by the net radiational loss to space. That is, the net radiation for the entire earth over a few years is very nearly zero" (Suomi 1957). Therefore, if we set $P_{emit} = P_{abs}$ as shown in Figure 9.1, the effective temperature of the Earth T_{eff} is given by

$$T_{eff} = \sqrt[4]{\frac{(1-\alpha)S}{4\varepsilon\sigma}}$$

where the variables and parameters are defined in the figure. The two key parameters needed to calculate the effective temperature of the Earth are therefore the Earth's emissivity and planetary albedo. While the public's perception of climate and climate change is mostly tied to greenhouse gases and consequently the changes of emissivity of Earth's atmosphere, the planetary albedo is of equal importance for climate (e.g., Stephens et al. 2015). And

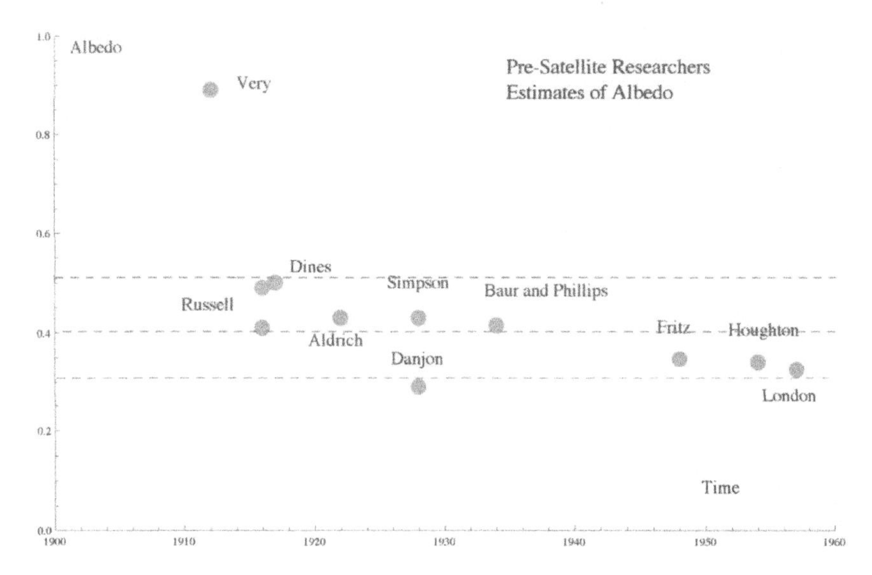

Figure 9.2. Pre-satellite researchers' estimates of albedo.

from a climate dynamics view, why do we care so much about the albedo? The papers by Wielicki et al. (2005) and Ramanathan (2008) have made it clear that a 0.01 change in the currently accepted value of albedo (0.29) has the same effect on global energy balance as a doubling of the current CO_2 concentration.

Pre-Satellite Estimates of Albedo

The stimulating article by Scripps Institution of Oceanography climatologist Ramanathan (2008) on the Earth's albedo is well worth revisiting in the context of Earth–atmosphere radiative equilibrium theory. Arguments in his paper highlight the sensitivity of albedo to several factors that govern it—especially cloud cover and cloud–aerosol interactions, where the aerosols are the nuclei for cloud drops. Indeed, it is a complex physical–chemical problem, and it is not surprising that estimates of the Earth's albedo in the pre-satellite age exhibited a wide range of values. Much of the uncertainty in these earlier studies stemmed from an assumed mean annual cloud distribution and the assumed albedo of cloud, typically taken to be 0.78 (now known to be significantly smaller). Further, Houghton (1954) conjectured that the early studies overestimated the water-vapor absorption. Hunt et al. (1986, Table 2) have presented these early estimates of albedo in tabular form, and we show these estimates in Figure 9.2. The latest pre-satellite estimates by meteorologists Sigmund Fritz (1949), Henry Houghton (1954), and Julius London (1957) were converging to a value in the range of 0.30–0.35.

The Path to Satellite Estimates of Albedo

Upon honorable discharge from the U.S. Air Force (USAF) in 1960, Frederick House (Figure 9.3) applied to the graduate program in meteorology at University of Wisconsin–Madison after reading about the program's active role in satellite meteorology. He was admitted to the graduate program in fall 1960 and assigned to work under Suomi. Both Suomi and Bob Parent had just begun to plan for an experiment aboard *TIROS-3*—the soon-to-be-launched third satellite in NASA's series of Television Infrared Observation Satellites (TIROS). *TIROS-3* was successfully launched on July 12, 1961, and operated for 230 days.

House began to examine data from *TIROS-3*. He recalled that "it was a period of excitement and real innovation to solve problems as they came up, especially as satellites were in orbit." The configuration of the radiometers on *TIROS-3* consisted of two sets of mirrors mounted on opposite sides of the satellite to maintain balance. A set consisted of one black hemisphere and one white hemisphere to discriminate between reflected shortwave radiation and emitted longwave radiation. A problem was identified where blackening of the white-paint pigment occurred due to absorbed ultraviolet radiation. This problem had also been noted on *Explorer VII*. This blackening introduced errors into the measurements.

House's dissertation (House 1965) focused on measurements from *TIROS-4*—launched on February 8, 1962, and it operated until June 30, 1962. Because of camera failure after June 10, House was only able to use data collected between February 8 and June 10 (122 days). House divided the dataset into segments, sequential 61-day groups labeled cycle 1 and cycle 2. He wanted to measure average radiative fluxes over separate two-month periods. That is, are there significant differences under the assumption of constancy in the instrument's function? As in the case with *TIROS-3*, the white radiometers blackened in time due to direct solar UV radiation. House's measurement strategy was that direct solar irradiance was measured twice each orbit during satellite sunrise and sunset. Reflected shortwave radiation was measured on the daytime side of the orbit. Simply put, the ratio of reflected shortwave radiation to direct solar radiation is the albedo. However, daytime observations include both longwave radiation as well as shortwave radiation from the Earth after direct shortwave radiation from the sum is subtracted from

Figure 9.3. Frederick B. House, 1965 (courtesy of the University of Wisconsin–Madison).

daytime observations. House found the following values of albedo during the cycles: cycle 1, 0.339; cycle 2, 0.355; and average, 0.347.

This result confirmed House's conjecture that there would likely be differences for the cycles—natural variability. At the time this work was completed (1965), little quantitative information existed about the angular distribution of reflected shortwave radiation and emitted longwave radiation. Information on this angular distribution is well known today (Wanner et al. 1997). Using this information, House (2015, personal communication) recalculated the albedo from these two cycles and obtained the following results: cycle 1, 0.317; cycle 2, 0.321; and average, 0.319.

In the early 1960s, there arose a major new opportunity to make radiation measurements from USAF satellites. Suomi was asked to assist the USAF in a highly secret satellite program (now declassified). It involved the use of a small satellite, a "scout or pathfinder," designed to determine cloud cover over a designated area that could then be used as input to plan activities for the Strategic Air Command. Suomi chose Thomas Vonder Haar (then a doctoral student under Suomi) as the civilian project leader. After assisting with the primary mission of this USAF project, the UW–Madison team was

Figure 9.4. Thomas H. Vonder Haar, 1968 (courtesy of the University of Wisconsin–Madison).

encouraged to develop and place other "experimental" instruments on the satellite. "Disc" radiometers, flat-plate radiometers (FPRs) similar to those on *TIROS-3* and *-4*, were placed on the USAF satellites.

The USAF satellites were placed in near-polar sun-synchronous orbits. This allowed the black and white FPRs to measure the entire global ERB, including the polar regions that had been missed by the lower inclination orbits of Explorer and TIROS series. The satellites' in-orbit configuration and stabilization acted as a "rolling barrel," with its spin axis perpendicular to the orbital plane (sometimes termed the wheel mode or cartwheel configuration). This allowed the FPRs to measure direct solar irradiance and be calibrated against it twice each orbit (Vonder Haar 1968). Virtually all ERB observations from the USAF program were successfully collected and preserved.

Suomi was ecstatic about results garnered from the global, all-season ERB data—obtained from the FPRs aboard USAF satellites. He was anxious to make these results known to the world's atmospheric science community. Vonder Haar (Figure 9.4) presented a paper on the ERB results at the 1968 American Geophysical Union (AGU) Fall Meeting in San Francisco, California, and Suomi simultaneously presented the results at the International

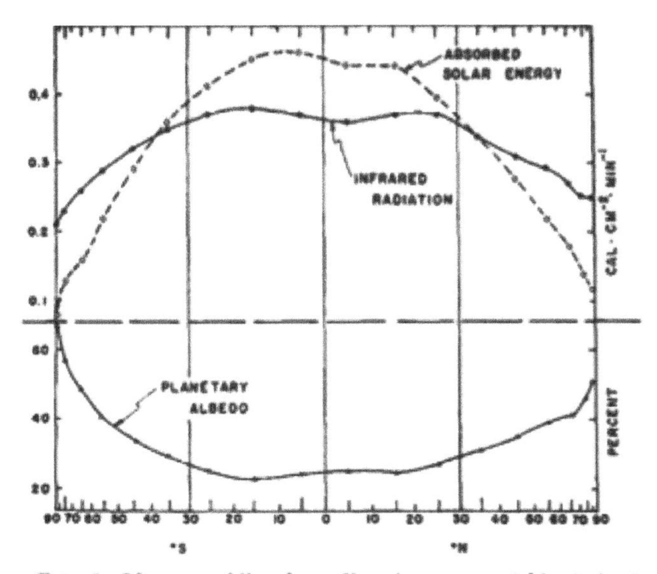

FIG. 1. Mean meridional profiles (averages within latitude zones) of components of the earth's radiation budget measured during the period 1962–66. The abscissa is scaled by the cosine of latitude.

Figure 9.5. Latitudinal dependence of average annual solar energy absorbed, thermal infrared energy emitted, and planetary albedo.

Radiation Symposium in Bergen, Norway. In summer 1969 both Suomi and Vonder Haar attended the Conference on the General Circulation of the Atmosphere in the Halls of the Royal Society, and shortly thereafter the new ERB results were published in *Science* (Vonder Haar and Suomi 1969)—and expanded upon two years later in the *Journal of the Atmospheric Sciences* (Vonder Haar and Suomi 1971).

The science community responded favorably to the presentations and publications, but questions arose since some results differed from earlier investigations. In particular, the Suomi–Vonder Haar results indicated that the Earth was "darker" than previously estimated (a lower albedo, 0.30 versus the earlier estimates closer to 0.35) and "warmer" (emitted infrared radiance higher by about 4%–5%). These results implied that the Earth–atmosphere system must accommodate more energy than previously believed. Importantly, results also indicated that each hemisphere has nearly the same albedo—somewhat surprising since the ocean–land contrasts in the hemispheres are significantly different. The essential results are captured in the now-famous graphic from Vonder Haar and Suomi (1971) that is shown in Figure 9.5—meridional averages of radiation and the associated distribution of albedo. These annual global results accurately identified the latitudinal bands where there is accumulation of heat (roughly between 30°N and 30°S) and diminution in the complementary region (from these latitudes to the poles). In short, oceanic and atmospheric heat transports between the regions of excess to the regions of diminution are required to maintain global heat balance. The variation in this annual average was further explored about ten years later, as found in Ellis et al. (1978).

Suomi and Vonder Haar remained active in ERB research through the 1970s and 1980s. Even into the last decade of Suomi's life, he remained active in ERB research—especially through advice delivered to the National Academy panels. This work spanned more than 40 years, and he declared it "mission accomplished" (Vonder Haar 2015, personal communication).

III

Space Science and Engineering Center: An Institute for Satellite Meteorology

CHAPTER TEN

Perfect Timing: NWP and Satellite Meteorology Merge

The Merger between NWP and Satellite Meteorology

In the 1950s, the meteorological world was abuzz with NWP and numerical experimentation that came with computational power. There was also some optimism about the possibility of using satellites to observe weather (Wexler 1954, 1960; Wexler and Neiburger, 1961). Three events occurred that melded these themes: 1) the feasibility of NWP (Charney 1948; Charney et al. 1950), 2) the numerical experiment conducted by Norman Phillips (Phillips 1956; Lewis 1998) that gave promise for extended-range forecasting (on the order of weeks), and 3) President Kennedy's call for the peaceful uses of space in his address to the United Nations General Assembly in 1961. Kennedy's speech fueled the adoption of UN Resolutions 1721 and 1802 (adopted in 1962) that led to the establishment of the World Weather Watch (WWW) in 1963.

The research arm of the WWW was the Global Atmospheric Research Program (GARP), collaboration between the International Council of Scientific Unions (ICSU) and the World Meteorological Organization (WMO). The Joint Organizing Committee (JOC) was formed and began deliberations to establish goals for GARP. Collection of global observations from space—central to defining the initial state of the deterministic extended range general circulation models—was among the primary goals [as retrospectively reviewed by Smagorinsky (1978) and Smagorinsky

Figure 10.1. Members of GARP's JOC are photographed at an October 1971 meeting in Downsview, Ontario, Canada. Others involved in the meeting are associated with the Committee on Space Research (COSPAR) and the International Council of Scientific Unions (ICSU). Standing (left to right): Fred Shuman, Morris Tepper, P. R. Pisharoty (India), Valentin Meleshko, Pierre Morel (France), Verner Suomi (United States), Fritz Möller (Federal Republic of Germany), John Sawyer (United Kingdom), Arnold Glaser, E. M. Dobryshman, Joseph Smagorinsky (United States), and Viktor Bugaev (Union of Soviet Socialist Republics). Sitting: Stanley Ruttenberg, Oliver Ashford, Bo Döös, Bert Bolin (Sweden), Robert Stewart, and Warren Godson.

and Phillips (1978)]. Suomi was ideally positioned to take the lead role in acquisition of global atmospheric observations and accordingly chosen to be a member of the JOC.

As stated by theoreticians Smagorinsky and Phillips, extended range deterministic forecasting by necessity must start with an accurate state of the atmosphere extending from the surface to the upper reaches of the troposphere and the globe (land, sea, and poles). And there was a solid theoretical basis for estimating thermodynamic structure through radiometric measurements aboard satellites—a scientific arena where Suomi and Parent were viewed as the pioneers. In the context of global observing systems, the potential impact of temperature structure (soundings) for the southern hemisphere was especially significant in view of its severe data voids. The theoretical basis rested on Lewis Kaplan's monumental work that linked satellite measurements of absorbed and emitted infrared radiation from polyatomic gases in the atmosphere to the vertical distribution of temperature along the line-of-sight beneath the satellite (Kaplan 1959)—expanded upon in chapter 15.

As mentioned above, Suomi joined an elite group of 12 meteorologists and atmospheric scientists from around the world as a member of the JOC

(Figure 10.1). Suomi's forceful argumentation and involvement in the JOC meetings is remembered by his close colleague Pierre Morel:

> I was the token Frenchman [in the JOC], an inexperienced thirty-something who had little to talk about. . . . Vern Suomi was the first who befriended me. . . . Vern Suomi stood out with a unique ability to bring rambling scientific or other arguments back on track. He was very effective in delivering his trademark exhortation to "fish or cut bait"—[stop vacillating and act on the subject at hand or disengage from it]—But for all his plain speaking, Vern was very easy to get along with, basically a very honest and likable person, precisely the mix of solid-based scientific expertise combined with deep-seated intellectual integrity (a rare commodity indeed in modern science society). (P. Morel 2007, personal communication)[1]

The respect that Suomi engendered from theoreticians is well expressed in Phillip Thompson's letter to the author (J. L.) as found in Figure 10.2. Suomi's international work was recognized in 1993 by the WMO when he was awarded the International Meteorological Organization (IMO) Prize (Figure 10.3).

Suomi's National Connections

In addition to his international connections, Suomi cultivated national relationships with senior scientific directors within NASA, NOAA (National Oceanic and Atmospheric Administration), and the USAF. These directors were the key decision-makers regarding financial support for scientific research related to meteorology, atmospheric science, and space science. Suomi had an especially strong connection with David Johnson (Figure 10.4), a highly respected director within the national satellite community—first as director of U.S. Weather Bureau's (USWB) Meteorological Satellite Laboratory in 1960 (coinciding with the launch of *TIROS-1*) and ultimately as director of the National Environmental Satellite, Data, and Information Service (NESDIS). Several of Suomi's early contracts with NESDIS and its forerunners were formalized with a handshake between these two men, an indication of their mutual trust. The basis for this trust is remembered by Paul Menzel: "In 1975, when I went to a meeting in Washington, it was

1. Morel's more complete evaluation of Suomi is found in the Vignettes section of the Appendix.

NATIONAL CENTER FOR ATMOSPHERIC RESEARCH

P. O. Box 3000 • Boulder, Colorado 80307

Telephone: (303) 494-5151 • TWX: 910-940-3245 • Telex: 45 694 • FTS: 322-5151

18 December 1981

Dr. John Lewis
Space Science Center, Rm. 219
1225 Dayton Street
Madison, WI 53706

Dear John:

Many thanks for your letter of 11 December and attached notes. As anticipated, the results clearly depend on the ratio of the wavelength to the space increment, as well as on the Courant number, $c\frac{\Delta t}{\Delta x}$. One, of course, hopes that Δt is small enoughthat the Courant number is appreciably less than unity.

I look forward to hearing the results of your new experiments, particularly those involving multiple time levels. You may be temporarily embarrassed by Suomi's enthusiasm, but in the long run it won't hurt. This is one of his ways to spur people on to greater heights, but he wouldn't do so unless he thought the basic idea a good one and worth pursuing.

I appreciate being posted and hope to hear from you soon. With best regards,

Sincerely yours,

Phil T.

Philip Thompson

The National Center for Atmospheric Research is Operated by the University Corporation for Atmospheric Research under sponsorship of the National Science Foundation.

An Equal Opportunity/Affirmative Action Employer

Figure 10.2. Letter from Phil Thompson extolling Suomi's ability to determine worth of scientific projects.

understood and sometimes stated explicitly, that Suomi with a couple of students could outperform a government lab because you'd get more 'bang for your buck.' Students are not expensive, there're on the steep part of the learning curve; he was able to continue to get contracts and grants because of this efficient and cost-effective performance on research."

Figure 10.3. Zou Jingmeng (left), President of the WMO, awards the IMO Prize to Suomi.

Figure 10.4. David S. Johnson (courtesy of NOAA).

Creation of SSEC

By the early 1960s, Suomi realized that his expansive dream of leading a scientific effort in satellite meteorology and space science would be difficult while serving as a professor, albeit a research professor, confined to the meteorology department at UW–Madison. He would need scientific teams, more than a talented colleague like Bob Parent, more than a set of graduate students, more than a single machinist or technician attached to the electrical engineering department. Suomi began to lobby for an institute at the highest levels within the university. He had experienced significant positive interactions with Dean Ingraham in the College of Arts and Sciences as mentioned earlier, and he would come to have a champion in Robert Bock, a molecular biologist and future dean of the graduate school (from 1967 to 1989) (R. Fox 2015, personal communication).

The Schwerdtfeger Library at UW–Madison documents the steps leading to the creation of the Space Science and Engineering Center (SSEC). It officially came into existence on 20 August 1965 following a resolution by the University of Wisconsin Board of Regents. Suomi was named its director. It was his dream institute, a multidisciplinary institute—not attached to a single department but to a diverse set of departments within the university that at least had a modest interest in space science. As remembered by Menzel, "he [Suomi] was early in favoring a multi-disciplinary approach. He would go over and get physicists, geologists, whoever came along. He would have a way of including the disciplines in the dialogue." The 15-story building that houses SSEC was completed in 1968, funded equally by the National Science Foundation, NASA, and the State of Wisconsin.

CHAPTER ELEVEN

Panoramic View of Suomi's Research Themes at SSEC

We first present a wide-angled view of research themes that dominated Suomi's scientific life. In this chapter, we borrow heavily from Paul Menzel's tribute to Suomi in the form of an expanded obituary published in AGU's newsletter *Eos* (Menzel 1995).

As stated earlier, Verner Suomi's major contributions to space science engineering began in the late 1950s when he and Robert Parent developed the flat-plate radiometer that measured the Earth's heat budget from a satellite. As discussed in chapters 7 and 8, these radiometers are impressively simple and free from large error; they can be calibrated by viewing the Sun, space, and Earth in sequence from a spinning satellite. These radiometers flew on the TIROS, ITOS (Improved TIROS Operational Satellite), and DMSP (Defense Meteorological Satellite Program) series of satellites; the earliest version was flown on *Explorer VII*. In Figure 11.1, outgoing longwave radiation measurements collected aboard *Explorer VII* on December 2, 1959, are overlaid on the surface pressure pattern and 500-hPa temperature field for that same day. Here we note a southwest-to-northeast swath of outgoing longwave radiation that exhibits low values over the Pacific Ocean that reach a maximum over the western United States in association with a surface anticyclone—typically associated with subsidence and an absence of cloud that leads to large values of outgoing radiation. From the western United States to

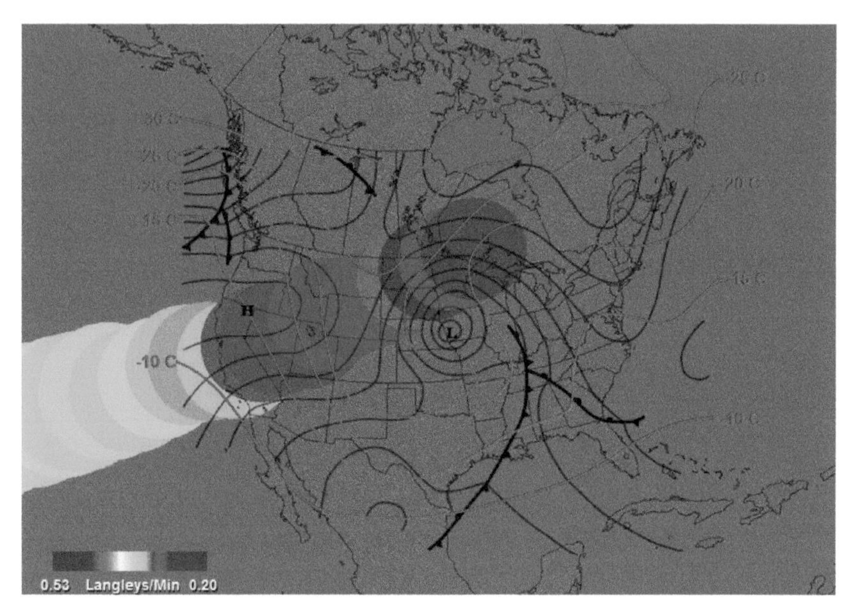

Figure 11.1. *Explorer VII* radiance measurements overlaid on surface pressure and 500 hPa temperature map. Radiances were found to increase over land in clear skies.

the upper Midwest, the outgoing radiation gradually decreases in the presence of lowering surface pressure. These observations were first presented in July 1961 (Suomi 1961), and a summary appeared in the *IGY Bulletin* three months later (IGY 1961). A portion of the succinct summary follows:

> There appears to be clear indication that large-scale patterns of outward radiation flux . . . are related to the large-scale features of the weather. Meteorologists had expected a variation of outgoing radiation with latitude, but the preliminary results show variation of the outgoing long-wave radiation with longitude as well. . . . It is also apparent that these results re-emphasize the important role that clouds play in controlling outgoing radiation. (IGY 1961)

There has been a continual effort at SSEC to improve the flat-plate radiometers used in these early investigations. The following quote from the former director of SSEC, Hank Revercomb, clarifies Suomi's view of this uninterrupted effort:

> Suomi never looked back, always ahead. I found that surprising when I first started to work with Larry [Sromovsky] at SSEC in the mid-1970s. We were working on the design of a new Earth Radiation Budget Observing System,

> ERBOS as Suomi called it. While Vern, along with Bob Parent, had originated this type of scientific observation with his hemispheric sensors on *Explorer VII*, and built and flown flat-plate sensors ever since, he didn't tell us to read material related to the earlier work or to make use of the same type of sensors! In fact, he refused to support an effort at NASA Langley that proposed to use his original hemispheric concept. (H. Revercomb 2015, personal communication)

Indeed, a current project at SSEC that has built on Suomi's lasting desire to more accurately determine the ERB is titled CLARREO (Climate Absolute Radiance and Refractivity Observatory), a NASA Tier 1 project. SSEC's involvement concentrates on improving infrared high spectral resolution measurements for providing a true benchmark of the Earth's climate. It offers unprecedented accuracy in these measurements, but foremost among CLARREO's novelties is in-orbit verification of the measurement accuracy used to access climate change.

The spin-scan camera (Figure 11.2), introduced by Suomi and Robert Parent in 1963, represented a revolutionary milestone in satellite instrumentation. It is generally considered to be their most influential invention. Inspiration came to Suomi while watching instant replay during a Green Bay Packers football game (E. Suomi 2014, personal communication). In analogy with instant replay, Suomi started to think about how to capture observations of a single weather phenomenon at frequent intervals—in essence, make a movie of the weather.

To achieve the lofty goal of monitoring a weather system at frequent intervals, that is, creating a moving picture of the weather, profound design problems had to be overcome. The platform carrying the camera needed to remain stationary relative to the rotating Earth, that is, located 22,000 miles above the Earth in accord with the Newtonian laws of gravitational attraction. Further, the picture-taking process from a camera attached to a spinning satellite in geosynchronous orbit was the key design problem. Homer Newell, director of NASA's Office of Space Science, was starting to think about this process in the 1960s, but Suomi and Parent were ahead of him (P. Menzel 2015, personal communication). Suomi and Parent used a scanning process that produced a full-disk Earth visible image every 23 minutes by piecing together separate strips of imagery collected as the satellite spun in orbit—1,821 separate images. Of course, the full-disk image was a hemispheric image viewed from a fixed longitude on Earth (Suomi and Parent 1964). On December 7, 1966, the spin-scan camera was placed on the

Figure 11.2. Spin-Scan Cloud Camera (courtesy Smithsonian Institution Archives, Image #7B10655).

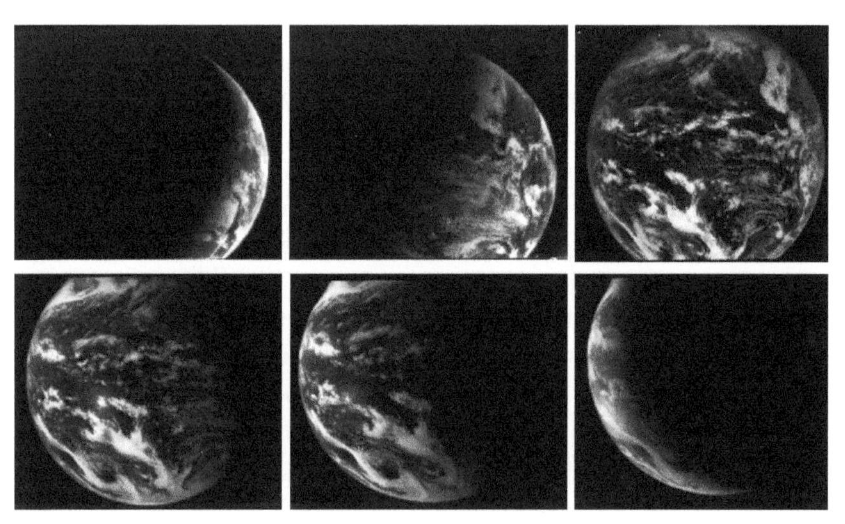

Figure 11.3. *ATS-1* Images of Earth, morning to evening on December 12, 1966 (courtesy UW–Madison Space Science and Engineering Center and NASA).

Figure 11.4. One of the color images of Earth taken from *ATS-3* on November 18, 1967 (courtesy UW–Madison Space Science and Engineering Center and NASA).

Applications Technology Satellite (*ATS-1*), and a series of photographs taken a few days later is shown in Figure 11.3. *ATS-1* was positioned over Earth longitude 80°W, and accordingly, the full-disk image covers the Western Hemisphere. The final product led Suomi to coin the quote of the moment: "now the weather moves, not the satellite." One of the color images from *ATS-3* is shown in Figure 11.4.

Two operational systems were based on the spin-scan design (Suomi and Krauss 1978): NASA's Synchronous Meteorological Satellite (SMS) and

NOAA's Geosynchronous Operational Environmental Satellite (GOES). The camera has also been adopted by the European Space Agency and the Japanese Meteorological Agency for their operational meteorological satellite programs. The development and application of a global geostationary imaging capability has had many positive consequences. Among the most important has been the timely alerts and warnings issued in anticipation of hurricane and typhoon landfall. Further, this camera has made meteorological satellite data routinely available to many nations that otherwise could not afford it, and it has extended the period of useful weather prediction (see chapter 15). The Visible Infrared Spin-Scan Radiometer (VISSR) Atmospheric Sounder (VAS) Experiment began in 1971 and was proposed by Suomi to sound the atmosphere's temperature and water vapor distribution from a geostationary satellite (NASA 1985). This majestic advance in satellite meteorology was accomplished through a team effort at SSEC involving engineers, technicians, and atmospheric physicists. The details of its development are explored in chapter 12. From design inception to operational implementation took 10 years (early 1970s to early 1980s). Among the most challenging aspects of this project was amplification of the low levels of terrestrial energy received at such high altitudes within the instrument's small field of view. A spirited celebration took place at SSEC when VAS's successful performance was demonstrated shortly after it was launched on *GOES-4* in September 1980 (Smith et al. 1981).

To achieve rapid and versatile access to the spin-scan camera data, Suomi directed the development and evolution of the Man computer Interactive Data Access System (McIDAS). The system rapidly analyzes and interprets millions of satellite observations in conjunction with a variety of other observations and thereby enables the human eye to interpret the results. McIDAS was initially conceived as a way to produce accurate cloud motion measurements from ATS, SMS, and GOES satellite data. McIDAS evolved into a powerful data management tool.

The power of this tool was never made more public than when a series of tornadoes swept through Kansas and destroyed major parts of two towns. As remembered by Don Johnson (Fox 2009, p. 7),

> a congressman from Kansas contacted NASA and wanted to know what they're doing because they'd made all these promises about weather. And what did they do? They turned around and called Suomi [and asked the question]: Can you help us in explaining to Congress what we're doing? So they actually, three of them came out here and we had a congressional hearing

> at the Space Science and Engineering Center. That led to the exportation of McIDAS's system that showed imagery, meteorological information to the Kansas City Severe Storms Lab [NOAA's National Severe Storms Forecast Center (NSSFC)], to Oklahoma [National Severe Storms Laboratory (NSSL)], to Washington and all over the world eventually. How many, I don't know.[1] They actually had a [McIDAS] system that observed the hurricanes, typhoons, I should say in the Bay of Bengal. . . . After McIDAS was in place, [a storm hit that area] and it happened that two or three hundred thousand people could have lost their life and the loss of life was down to five thousand. Suomi was very proud of that fact.

The McIDAS training of forecasters at the National Hurricane Center (NHC) is discussed in the context of atmospheric motion vectors in chapter 15.

Suomi's expansive view of weather and climate was never more apparent than when he started to focus on the atmospheres of neighboring planets in the solar system. This venture began in the early 1970s when he was appointed a member of NASA's Mars/Venus/Mercury Imaging Science Team and the Jupiter/Saturn Imaging Team. His membership on these teams was followed by service to NASA as a member of their Advisory Committee on Scientific Uses of Space Stations. In chapter 14, we investigate the range of scientific activities that surround this work on the atmospheres of neighboring planets.

Another far-reaching application of satellite data was use of radiances from the infrared "window," the Earth–atmosphere radiation emanating from the spectral region close to 11 μm. At this location in the spectrum, infrared radiation is transmitted through the clear-sky atmosphere without any significant absorption, and thus is indicative of temperature changes caused by the presence of clouds, land versus sea temperature differences, and atmospheric moisture variations. Therefore, the "window" sensor enables assignment of cloud heights and determination of cloud tracks at night, both important for NWP. In the presence of clouds, the lower atmosphere of Earth is obscured and the radiation emanating from the colder clouds provides a contrast to that emanating from the warmer Earth surface. Overall, the infrared window images reveal the location of clouds, surface features such as snow and ice cover, frozen versus open water, and low-level atmospheric moisture accumulations.

---

1. Tom Haig indicated that upwards of 60 McIDAS units were distributed around the world during the time he served as executive director of SSEC (1970–1980).

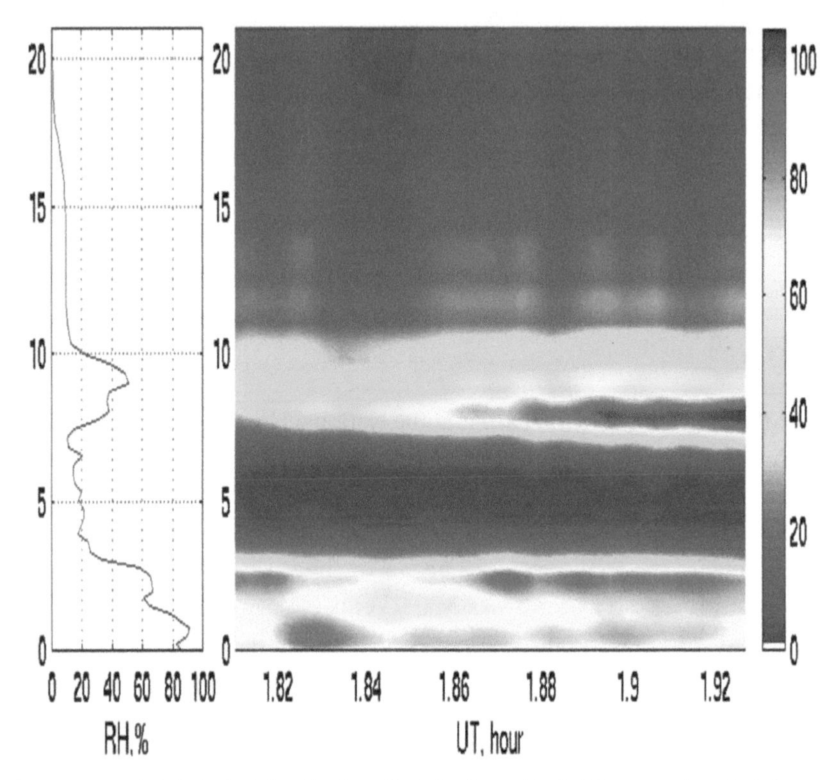

Figure 11.5. (right) Vertical distribution of relative humidity (RH) over a 1-h period that was determined from airborne infrared high spectral resolution measurements and (left) the RH profile from a nearby radiosonde at 0000 UTC.

Figure 11.6. A NASA depiction of *Suomi NPP*, taking cloud pictures from polar orbit.

Suomi's research ventures also included oceanography, especially as it related to the heat transport by oceans in response to radiative imbalances in the Earth–atmosphere system. His work in this area is covered in chapter 13. Further, his efforts to develop a sea-surface sonde, a device that was designed to measure the heat and moisture exchange at the air–sea boundary, an effort that consumed him to his last days, is presented in chapter 16.

In the 1980s, Suomi and Bill Smith collaborated to explore the information content of high spectral resolution infrared images, first from aircraft with the nadir-viewing High-Resolution Interferometer Sounder (HIS) (Revercomb et al. 1988) and later with a scanning version of the same instrument, the Scanning HIS. These measurements proved to be very useful for understanding and improving our knowledge of the basic spectroscopy and physics associated with various atmospheric processes, as well as making three-dimensional determinations of atmospheric moisture and temperature (see Figure 11.5). Later instruments provided global high spectral infrared measurements from polar-orbiting platforms, including the National Polar-Orbiting Operational Environmental Satellite System (NPOESS) Preparatory Project (NPP) satellite launched by NASA in 2011. On January 24, 2012, NASA gave this satellite a new name. They named it the *Suomi National Polar-Orbiting Partnership* (*Suomi NPP*). It flies with an impressive instrument package that is a testament to Suomi's lifelong work (Figure 11.6).

CHAPTER TWELVE

Suomi's Model for Conducting Research at SSEC

The SSEC, a scientific institute within a major research university, continues to follow Suomi's approach to business 20 years after his passing. Although the workforce at SSEC is now large, the prototype scientific team consisted of a scientist and engineer—Suomi and Parent. The team size had the luxury of increasing as SSEC grew, but there was always a tendency to avoid overlap among the members—teams as small as possible. Rather than outlining a recipe or method of attack on general research problems at SSEC, we follow the development of the VAS project, from its inception to operational implementation. We first summarize Suomi's style of leadership.

Management at SSEC

It is interesting to compare the scientific-administrative styles of Suomi and his mentor Carl Rossby. Both, of course, were premier scientists drawn into scientific administration. Knowledge of their styles comes from interviews with their administrative assistants—Robert (Bob) Fox and Tom Haig in Suomi's case, and Horace Byers and Bert Bolin in Rossby's case.[1] Difficulties

1. Correspondence with Byers and Bolin took place in the 1990s when the author (J. L.) interviewed and corresponded with Rossby's associates (Lewis 1992). Interviews with Fox and Haig took place in 2015.

in the day-to-day handling of institute affairs arose in response to their frequent and extended travels—they continually needed to search for institute funding, and they were central to decisions made on the world's scientific stage.

We restrict our assessment of SSEC's management to the period between 1970 and 1995—the period when Tom Haig served as executive director of SSEC (1970–1980) and when his successor Bob Fox served as executive director (1980–1995). Photos of Suomi in the company of these two executives are shown in Figure 12.1 (Suomi and Haig) and Figure 12.2 (Suomi and Fox). The relationship of Haig to Suomi was in the mode of colleagues. They had collaborated on military and national satellite programs in the 1960s, especially those that emanated from the National Reconnaissance Office (NRO) where Haig, then a USAF colonel, served as a key scientific officer. On paper, the union between Suomi and Haig appeared ideal, but their personalities sometimes clashed. Among the differences of opinion were the following:

1. Haig's handling of SSEC correspondence during Suomi's extended absences—Suomi wanted no meddling in his affairs despite the need for timely institute decisions while he was absent from SSEC;
2. Haig's interaction with the School of Computer Science and the decision to purchase a computer for SSEC during one of Suomi's long absences from the institute; and
3. the absence of a clear-cut policy for interaction between project managers and Haig—not infrequently, Haig was reprimanded by Suomi for decisions made between himself and a project manager.

Nonetheless, Suomi and Haig ushered in many new programs at SSEC, including McIDAS and the VAS Demonstration Project.

Fox was a strict military man (USAF) who had served in high-level management positions before he assumed the executive director position at SSEC. He was also a very capable meteorologist who received his PhD at UW–Madison under Professor James Weinman. Although soft-spoken and most courteous in conversation with those under his supervision, he exhibited an ordered manner of conducting business and seemed to have a set of rules that governed his actions. It appears that Suomi broke at least one of Fox's rules in their first conversation that took place in early 1980. Fox's memory of the conversation follows (R. Fox 2015, personal communication):

Figure 12.1. Tom Haig (left) and Verner Suomi next to an early interactive meteorological processing and display system in 1973 (courtesy of Edwin Stein, *Wisconsin State Journal*).

Figure 12.2. Verner Suomi and Robert Fox (courtesy of SSEC, UW–Madison).

> It was my first day of work and Suomi visits me. I didn't know Suomi well except as a student years earlier. Don Johnson [Professor of Meteorology at UW–Madison and a trusted colleague of Suomi] was my contact in the hiring process. Suomi says 'What can I do to get you oriented.' I told him that wherever I'd worked, I liked to look at the 5- or 10-year plan for the organization and wanted to know the long-term objectives. There was a long silence. Then Suomi said, 'Your job is to go where I want to go when I decide I want to go there, and keep me legal and out of trouble. And I might change direction half way through the day.' He then walked away. I later found there was no 10-year plan, let alone a 5-year plan, there wasn't even a 1-year plan.

Despite this puzzling discussion on Fox's first day at work, these men came to greatly respect and trust each other (R. Fox 2015, personal communication). Both came from blue-collar backgrounds and shared a respect for the workingman and workingwoman. Fox is remembered as the executive who adroitly handled the scientific project funds and assured the workforce of continued employment even though transfers from one project to another project were standard practice.

Fox recalls how Suomi viewed the SSEC budget: "It's far easier to get ten million dollars in 5 years than getting one million dollars next year. Plan. Look to the future." Fox continued: "And if you remember how he did things, he worked on concepts and always looked to the future. He was always putting things in the budget without straining today and next year's budget. He watched that and I firmly agreed with him."

An Example of Problem Solving: The VAS Project

The Idea: VAS

The name VAS is an acronym within an acronym: VISSR (Visible and Infrared Spin-Scan Radiometer) Atmospheric Sounder and the project was central to SSEC's activities from the early to mid-1970s to the mid-1980s. VISSR was an imaging instrument aboard NOAA's *SMS-1* satellite, the first operational satellite designed to sense meteorological conditions in geostationary (geosynchronous) orbit over a fixed location on Earth. The spin-scan radiometer had evolved from the first Suomi–Parent camera that initiated geostationary viewing of the Earth (discussed earlier in chapter 11).

Suomi believed it possible to make VISSR a dual-purpose instrument—an imager, as the instrument was originally designed, and a "sounder," an instrument that could measure/estimate the atmospheric temperature and moisture profile at fixed locations beneath the satellite. If this could be accomplished, then the visible and infrared imagery could be complemented by the quantitative information on atmospheric structure. Indeed, the weather system could then be followed in a Lagrangian frame of reference (a coordinate system that moves with the "weather"—with the air parcels).

A myriad set of questions arose regarding collection of radiances from the Earth and its atmosphere at a distance of ~35,000 km, the equilibrium position of a satellite in geosynchronous orbit about Earth. Compared to the polar-orbiting satellite, the distance factor is approximately 45 times greater. And energy receipt at the instrument site goes as the inverse of squared distance, so the ratio of radiant energy received at geosynchronous altitude compared to energy received at polar-orbiting altitude is 45^{-2} ($=5 \cdot 10^{-4}$). To overcome the relatively weak signal at geostationary altitude, multiple measurements of the radiance are required to reduce the noise in observations—essentially averaging over many soundings in a process labeled "dwell sounding." Further, a partitioning of time on VISSR to both imaging and sounding had to be devised.

The Filters

Suomi was confident that his dream of following weather from geostationary altitude was possible, but he needed input from a cadre of youthful physicists who had joined him at SSEC—Larry Sromovsky, Hank Revercomb, and Paul Menzel, all recent PhDs in theoretical physics from UW–Madison's physics department. They served as "filters" for Suomi—trusted colleagues, indeed youthful colleagues, who listened to him and were unafraid to disagree. Here they discuss this subject of "filters," where name abbreviations LS, HR, and PM are used to distinguish their commentary:

> [HR]: Suomi had his strengths for sure, but he didn't know everything and he knew he didn't know everything. How did he get around this uncertainty? Well, he had *filters*—people he trusted, who knew more about a particular topic than he did. They weren't afraid to disagree. . . . Larry Sromovsky has a keen understanding of physics and good intuition, and he served as a filter enormously. . . . He [Suomi] didn't like it if you told him it was a bad idea, but he did appreciate it in the sense that he would move on to a better idea or he would persist if he thought you were wrong. . . . He knew how to use his people.

> [PM]: Suomi's life was full of those things where he had confidence, but he realized his shortcomings; in moments of honesty, he would say something like that.... He had those sanity check guys, because he didn't totally trust the details. But his big vision he never questioned. He knew what the questions were, he knew where you could find the answers.

> [LS]: Well, the filters weren't as good as they should have been. . . . On so many of the projects, they were new and we had no experience, so we weren't so good . . . it was only the second or third time you did something, like with the radiometers, that you could say "good."

Charles "Chuck" Stearns (a doctoral student under Suomi and later a professor at UW–Madison) validated this last point made by Sromovsky. Stearns was Suomi's "right-hand grad student" who assisted him on work with the ping-pong radiometer (Figure 7.3). In his recollection, Stearns said, "we were doing things we had never done before . . . we had to have short term memory for the data [and we used] a little tape recorder. You could wind it up on a spring driven reel and let'er go and have it play back and go down to the ground [station]. And it worked" (quotation in Fox 2009). The idea for this spring-loaded device came from Bob Parent, "and it was just a magnificent little device that had the look of one of those large springs [mainsprings] in a 'Big Ben' alarm clock" (T. Haig 2015, personal communication).

The Execution

Stage I of the VAS project centered on building the radiometer, developing in-flight calibration, demonstrating noise reduction through averaging the radiances, and the determination of sounder event timing. Suomi succeeded in garnering NASA money for Stage I (NASA Contract NAS5-21607). The final report was titled "SMS Sounder Specification." One of the most impressive results in the final report centered on application of radiometric soundings to 1) a tornado over northeast Wisconsin and 2) Hurricane Ginger over the Atlantic on the same day—28 September 1971. Simulated soundings in clear air adjoining the disturbances were available at intervals of 1–2 minutes with horizontal resolution of ~15 km. The following statement appeared in the report: "Thus the geostationary sounder presented in this document could represent an enormous step forward in our ability to observe small-scale severe weather phenomena." The statement needed some qualification: the simulated observations were only available in clear-air regions.

Despite these limitations, Suomi felt confident that a second phase of the VAS project was justified—a phase that would "demonstrate" the ability to collect and process these observations in a "near operational" time frame. Thus, in November 1975, Suomi organized a meeting to discuss Stage II of the VAS project. The meeting was held at SSEC, and the NASA review team included Drs. James Dodge and Michael Garbacz. Paul Menzel remembers preparation for the meeting as follows:

> The NASA visitors anticipated a review of the ground system. In short, the system consisted of a NASA instrument modification [VISSR modification] on a NOAA platform [SMS] that is going to be processed at University of Wisconsin. The review was going far beyond the ground system. Vern was pretty tense in anticipation of this review and he wanted a "dry run," unusual for him. A big part of the center's future was resting on support of VAS. I was tasked with talking about the hardware, Hank and Larry would talk about the different modes of operating the instrument . . . instrument details. (P. Menzel 2015, personal communication)

A discussion of Suomi's presentation is complemented by the viewgraphs he used (labeled VS1 through VS5 as found in Figures 12.3 and 12.4). Vern's presentation focused on the end goals, not the instrument as captured by the phrase on VS1: "not a gadget but a program." VS2 erases the thought that hardware is the centerpiece. The statement "they buy it early" is indicative of a program goal to buy hardware later, when it is cheaper. Vern's drawing in VS3 again stresses the point that VAS is software, not only hardware. The "VAS ship" steams along and sees the "tip of the iceberg," evidence of software, but much more is hidden from view (below the iceberg tip).

VS5 hits close to the target of VAS: get rid of "gaposis," Vern's snappy word that describes the ubiquitous problem of huge gaps or data voids in atmospheric observations. Note that he has emphasized going from the large scale to the smaller scale in concert with the results discussed in the Stage I final report. His last graphic (VS6) presents a view of the milestones on the path to this new venture. Somewhat hidden, but most important, is Suomi's plan to bring Bill Smith (a former Suomi protégé—see the Vignettes section in the Appendix) and his NOAA/NESS (National Environmental Satellite Service, the precursor of NESDIS) group to Madison to be a part of the VAS project. Smith and his team were leaders in software development for production of temperature profiles from satellite radiance measurements. In fact, a decision had already been made to move Smith and his group to SSEC before

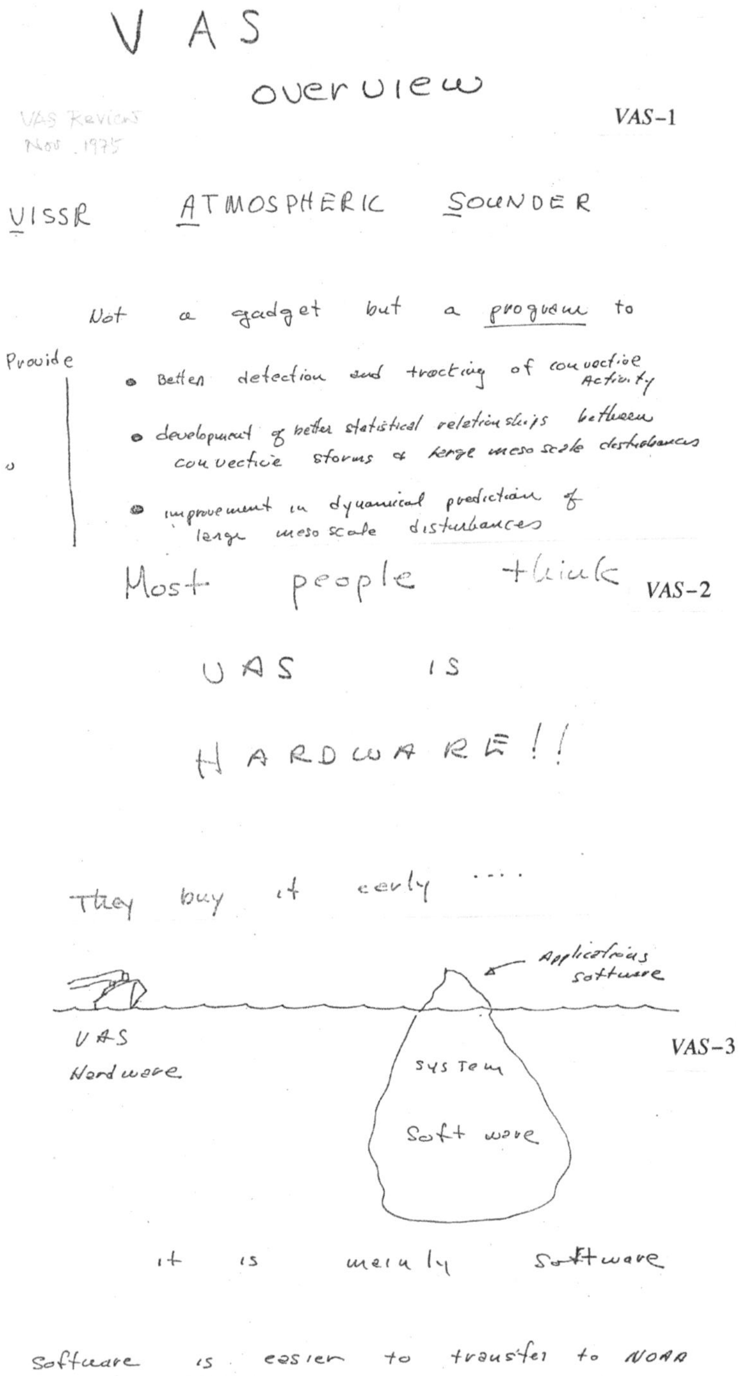

VAS

overview

VAS Review
Nov. 1975

VAS-1

VISSR ATMOSPHERIC SOUNDER

Not a gadget but a program to

Provide

- Better detection and tracking of convective activity
- development of better statistical relationships between convective storms & large meso scale disturbances
- improvement in dynamical prediction of large meso scale disturbances

Most people think

VAS-2

VAS is

HARDWARE!!

They buy it early ....

Applications software

VAS Hardware

System Software

VAS-3

it is mainly software

software is easier to transfer to NOAA

Figure 12.3. Suomi's notes (VAS1–VAS3) for NASA review.

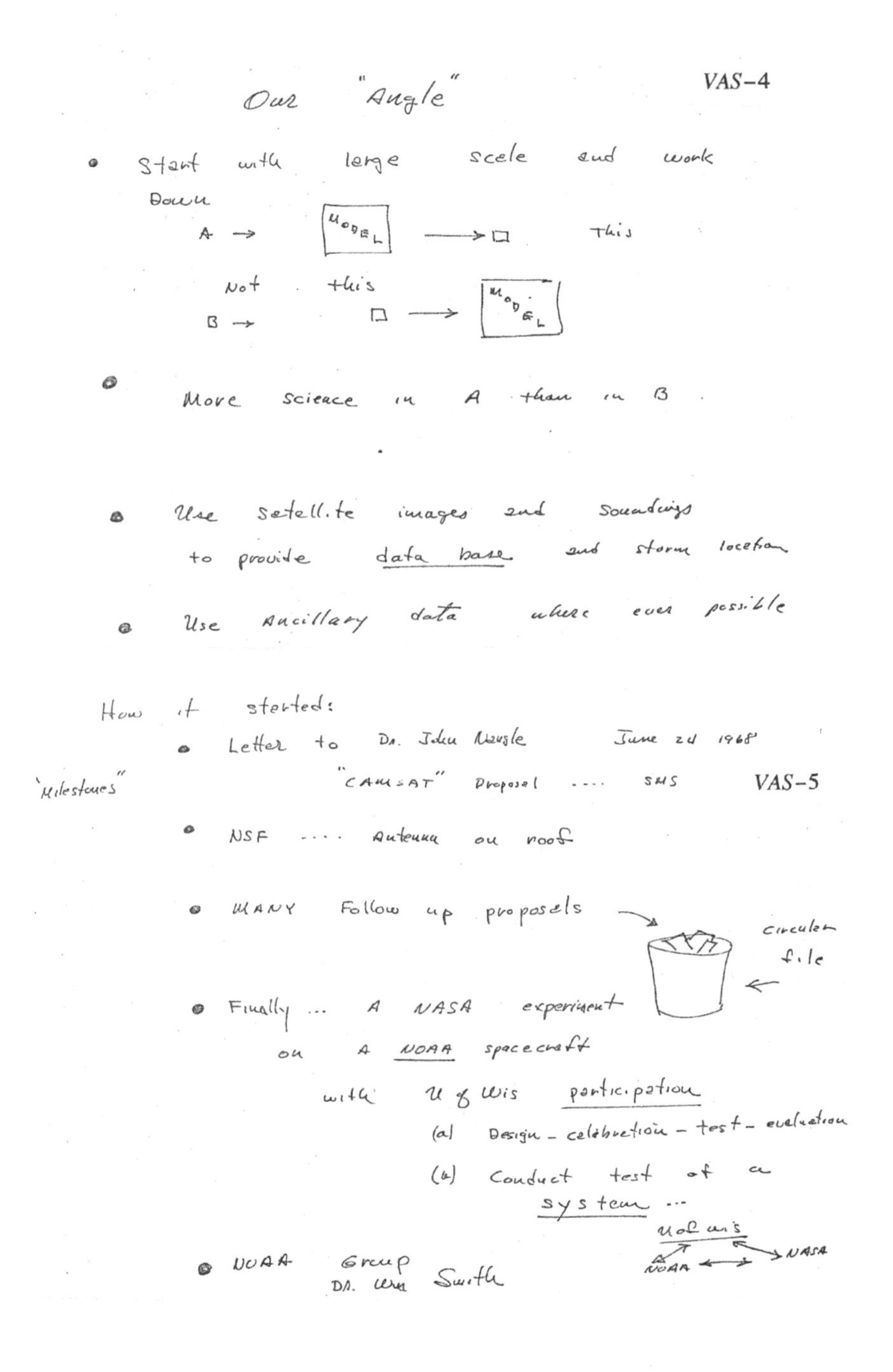

Figure 12.4. Suomi's notes (VAS4–VAS5) for NASA review.

Figure 12.5. Bill Smith's NOAA/NESDIS group. Back row, from left: Leroy Herman, Frederick Nagle, John Lewis, Hugh (Ben) Howell, Geary Callan, William Togstad. Front row, from left: Harold Woolf, William Smith, Christopher Hayden (courtesy of the Cooperative Institute for Meteorological Satellite Studies, UW–Madison).

the high-level VAS meeting took place—a decision made by NESS director Dave Johnson. Several members arrived at SSEC in 1976, and a picture of all members of the group is shown in Figure 12.5.

Suomi's Stage II of VAS was funded under the name VAS Demonstration Project. The project began in October 1980. The veracity of the VAS demonstration—ability to produce images and soundings in near–real time from geostationary altitude—came from results presented in Smith et al. (1981). Suomi's dream had come true. In his excitement at seeing the outstanding formal reviews of the paper, Suomi asked Smith if a flip of the coin might be appropriate to determine the lead authorship of the paper. Bill said, "No thanks, Vern, I think I would lose!" (P. Menzel 2015, personal communication).

Applicability of the VAS products to severe storm forecasting was hampered by the poor vertical resolution of the soundings. The detailed thermodynamic structure in the lowest levels of the atmosphere—typically below 850 hPa—could not be resolved. Bill Togstad, a member of Smith's NOAA/NESDIS team with experience as a National Weather Service forecaster, was tasked with the job of evaluating VAS in severe-weather situations. He remembers this work as follows:

> [at the time, the early 1980s] I could not in good conscience say that the satellite VAS product materially helped diagnose severe weather onset; perhaps it was my lack of imagination at the time, because Dr. Ralph Peterson did find a valid positive contribution in the form of showing areas of convective instability formation which, as any good severe weather diagnostician knows, is instrumental in some of the most violent severe weather outbreaks. (W. Togstad 2015, personal communication)

Despite this initial disappointment, the work of Lewis and Derber (1985) made it clear that the "gaposis" ailment could be ameliorated with VAS. They were able to use geopotential fields produced by VAS at 3-h/25–50-km resolution to follow the development of waves in a large-scale baroclinic weather system over the central United States. By the mid-1980s, a series of publications by Smith and colleagues placed VAS products in a strong position for operational implementation (Smith et al. 1981, 1982; Menzel et al. 1981, 1983a,b). VAS continued in a demonstration mode until geostationary soundings became operational with the introduction of *GOES-8* in 1994.

IV

Notable Research Themes: Their Past, Present, and Future

CHAPTER THIRTEEN

Ocean–Atmosphere Interaction

Vern Suomi learned of the role played by oceanography in the Earth's climate in his graduate studies at Chicago. It was through classes, seminars, and interactions with Rossby and Starr that this connection was firmly established. Indeed, Suomi often encouraged his students to read Rossby's memorial volume, *The Atmosphere and the Sea in Motion* (Bolin 1959), which contains papers dealing with oceanography as well as numerous papers on meteorology. Rossby's development as a research scientist was closely tied to his connection with oceanographers at the Woods Hole Oceanographic Institution from the 1930s through the 1950s (Lewis 1996). As remembered by noted Woods Hole oceanographer Henry Stommel, "Rossby was intensely interested in what he was doing. He'd ask people to come to his rented house at Woods Hole in the evening—and we'd stay till after midnight working out equations, etc., at the kitchen table." A photograph of Stommel is shown in Figure 13.1, and his letter to the author (J. L.) that contains the quote is found in Figure 13.2.

As the ERB datasets increased in size, the average annual cycle of energy exchange between Earth and space could be calculated with some fidelity. One of the first noteworthy results came from calculation of annual and seasonal poleward energy transport by atmosphere and ocean in combination—transport required to maintain global heat balance (Vonder Haar and

Figure 13.1. From left to right: Nora Charney (Jule Charney's daughter), Jule Charney, and Henry Stommel, on the occasion of honorary doctorates given to Stommel and J. Charney at the University of Chicago (1970) (courtesy of G. Platzman).

Suomi 1969, 1971). Suomi and his students were excited to find that 40% more transport was required to maintain balance than previously estimated—estimates made in the pre-satellite age. The implications of this result were substantial. It meant that earlier work had underestimated the strength of the atmospheric and, perhaps, the oceanic circulation systems. If, as suspected, the ocean was moving more heat toward one or both polar regions, then the exchange of energy between air and ocean was more important than previously thought.

Suomi and his colleague Joseph Smagorinsky, the founding director of NOAA's Geophysical Fluid Dynamics Laboratory (GFDL), strongly encouraged Thomas Vonder Haar (faculty member at Colorado State University after receipt of his doctorate at UW–Madison in 1968) and Abraham Oort (GFDL scientist with a doctorate under V. P. Starr at MIT) to conduct research on this question of the relative roles of ocean and atmosphere in poleward energy transport. They accepted the challenge and presented their results in Vonder Haar and Oort (1973) and two subsequent papers published in the *Journal of Physical Oceanography* (Oort and Vonder Haar 1976; Carissimo et al. 1985). They discovered that the ocean was transporting approximately twice the energy as previously believed by oceanographers (Figures 2-14 and 13-1 in Hartmann 1994). This result caused quite a stir in the oceanographic community and contributed to awakening the interest of

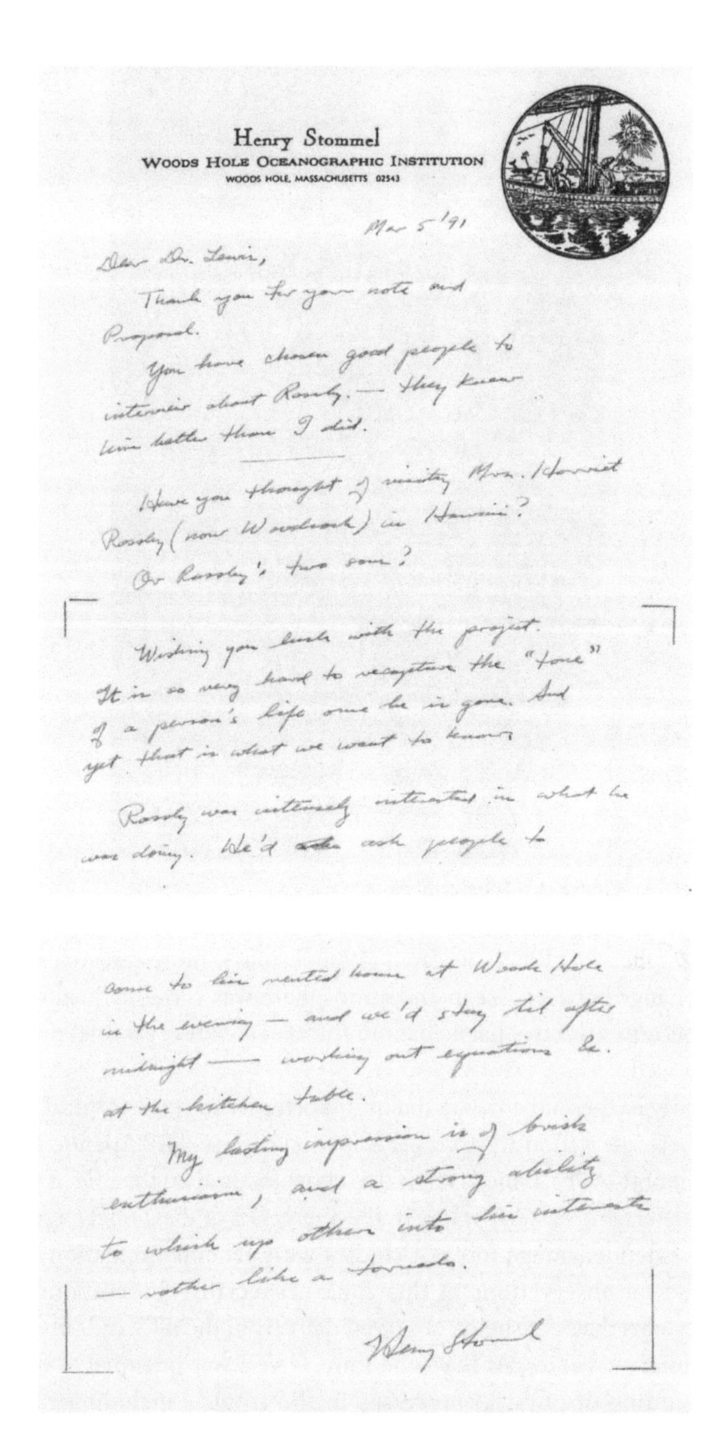

Henry Stommel
Woods Hole Oceanographic Institution
Woods Hole, Massachusetts 02543

Mar 5 '91

Dear Dr. Lewis,

Thank you for your note and proposal.

You have chosen good people to interview about Rossby.— they knew him better than I did.

Have you thought of visiting Mrs. Harriet Rossby (now Woodcock) in Hawaii?

Or Rossby's two sons?

Wishing you luck with the project. It is so very hard to recapture the "tone" of a person's life once he is gone. And yet that is what we want to know.

Rossby was intensely interested in what he was doing. We'd ~~ask~~ ask people to come to his rented house at Woods Hole in the evening — and we'd stay til after midnight — working out equations &c. at the kitchen table.

My lasting impression is of brisk enthusiasm, and a strong ability to whisk up others into his interests — rather like a tornado.

Henry Stommel

Figure 13.2. Letter from Henry Stommel describing Rossby's association with Woods Hole Oceanographic Institution.

Figure 13.3. Suomi and Arakawa.

oceanographers in climate research (Wunch 1997). More research into the heat exchange between ocean and atmosphere was certainly justified, and Suomi became an active participant in the oceanic observational programs that followed.

As early as the mid-1970s, a major opportunity to study tropical ocean–atmosphere interaction arose via participation in the GARP Atlantic Tropical Experiment (GATE). Objectives of this effort focused on the role of tropical cloud clusters and easterly waves in the energetics of the global circulation. GARP's extended-range forecast studies with general circulation models depended on observations in this zone—especially observations of the cloud systems that were being observed daily from the ATS and subsequent geostationary satellites. At the same time, GATE was intended to improve understanding of physical processes in the tropics, including radiative heating and cooling in the atmosphere and energy exchange between the atmosphere and ocean. This led to productive interactions between Suomi and two professors from UCLA, general circulation modeler Akio Arakawa and tropical meteorologist Michio Yanai. Through modeling and in situ upper-air observation of cumulus cloud clusters, these UCLA scientists were complementing Suomi's observational effort to understand the heat and moisture transport from the tropics to latitudes farther north and south. Suomi used cloud structure from satellite imagery and in situ measurements in the boundary layer to help analyze the transport. A photo of Suomi and Arakawa is shown in Figure 13.3.

Suomi and a team of SSEC colleagues developed the Boundary Layer Instrumentation System (BLIS), a novel array of instruments that could accurately measure the structure in the convective layer below cloud base. From tethered balloons aboard a ship, this system measured the 3-D wind and thermodynamic variables from the sea surface to 1500 meters. From these measurements, the energy and moisture fluxes between ocean and atmosphere were calculated in support of GATE's boundary layer subprogram. SSEC scientists Don Wiley and David Martin worked closely with Suomi on this boundary layer effort using BLIS.

Research ships were deployed in a hexagonal configuration just north of the equator and west of Dakar, Senegal. BLIS contained five Boundary Layer Instrument Packages (BLIPs) hovering from a single tethered balloon. A BLIP is shown in Figures 13.4 and 13.5, and Figure 13.6 shows the instrument during testing (SSEC 1973; Phillips 2014).

Vern Suomi's interest in the complex interplay between ocean and atmosphere never diminished. One of his last major projects centered on an instrument package designed to rest on the turbulent ocean surface and measure flux of heat and moisture from sea to air and vice versa. It was called the Skin-Layer Ocean Heat Flux Instrument (SOHFI) (Figure 13.7), and it is discussed in some detail in chapter 16. He also had plans for a related instrument package called the Surface Contact Multi-sensor Float (SCMsF).

Like his mentor C.-G. Rossby, Suomi took a macroscopic view of the geophysical system that accounted for interaction between ocean, land, and atmosphere. In essence, both men realized that investigation of climate and equilibria of the Earth–atmosphere–ocean system required knowledge of the coupling between these components. Their students were introduced to the foundations of fluid mechanics and were encouraged to take part in field programs that explored "the atmosphere and the sea in motion." Suomi excelled at designing and developing sensors that could withstand the harsh environments of a turbulent sea and space far above the Earth's surface. A host of technicians and engineers as well as lead scientists were entrained

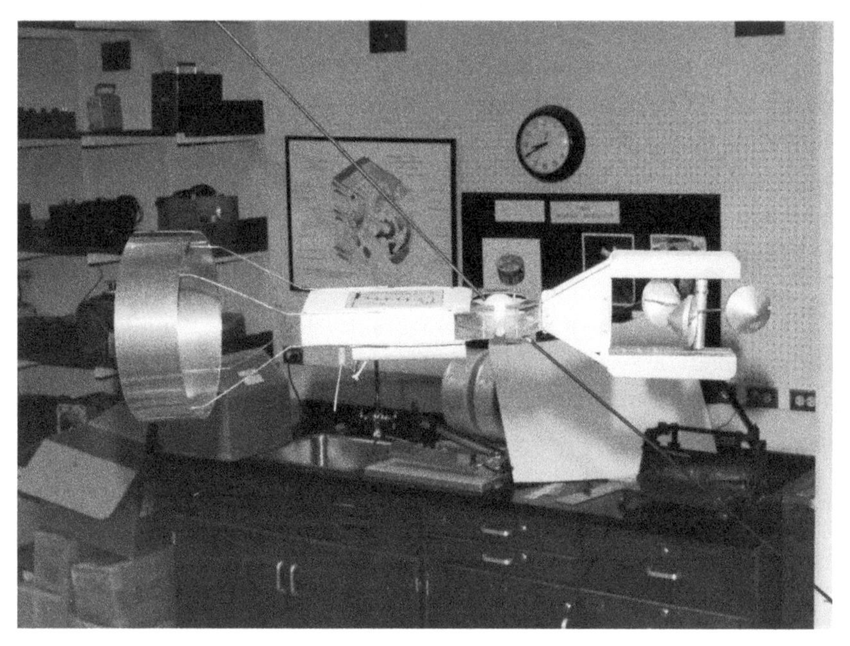

Figure 13.4. A tethered sonde, the BLIP, was designed to be used from ships during the GATE (courtesy of SSEC, UW–Madison).

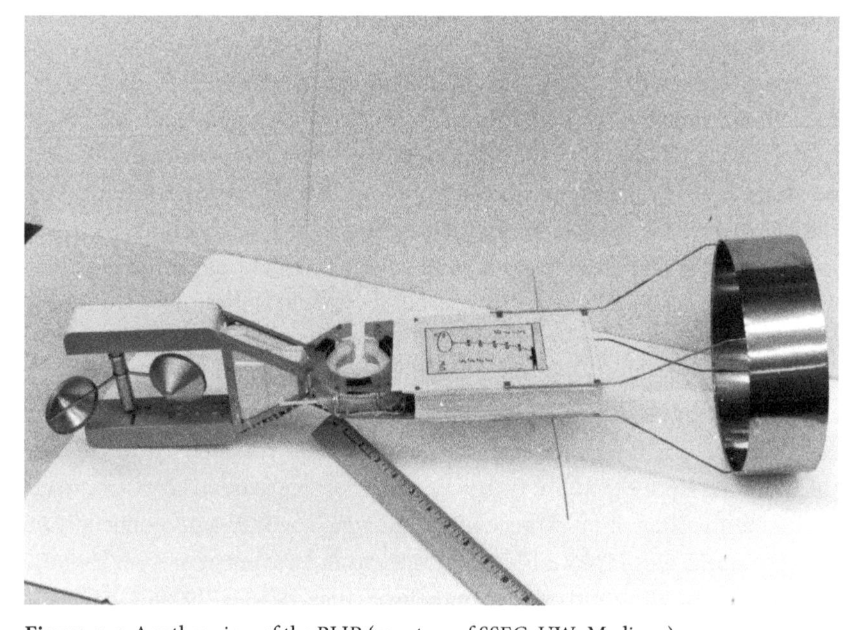

Figure 13.5. Another view of the BLIP (courtesy of SSEC, UW–Madison).

Figure 13.6. BLIS BLIP test [Portable Data Acquisition System (PODAS) stability runs] conducted from below radar antenna to F-deck stern aboard the ship *Discover*. Note the proximity to the ship's superstructure (courtesy of SSEC, UW–Madison, 1973).

Figure 13.7. Verner E. Suomi and Styrofoam prototype of the SOHFI he invented and patented (courtesy of University of Wisconsin–Madison Archives 1990).

into this exciting and expansive adventure. Among those who took a leading role in the work were Thomas Vonder Haar, David Martin, Don Wiley, Larry Sromovsky, Evan Richards, Jim Boyle, and Fred Best.

Suomi's research thrust has made it clear that he viewed the ocean and atmosphere as a conjoined or coupled system. Thus, it should not surprise us that the American Meteorological Society's annual Suomi Award captures this duality. The award is alternately presented to an oceanic scientist and an atmospheric scientist.

CHAPTER FOURTEEN

Atmospheres of Neighboring Planets

In the early 1960s, NASA and the Jet Propulsion Laboratory began to plan a series of missions to our neighboring planets—Venus and Mercury toward the Sun (relative to Earth) and Mars away from the Sun. The ambitious project assumed the name Mariner, where increasing positive integers following this name identified sequential missions. The first successful launch occurred with *Mariner 2* on August 22, 1962—a three-and-a-half-month flight to Venus. Images of the planet were obtained when the spacecraft passed the planet—on the "flyby"—and thereafter came into orbit about the Sun. Each Mariner mission had specific objectives, ranging from a close look at Mars (*Mariner 4*) to investigation of the chemical composition of Venus's atmosphere (*Mariner 5*), and *Mariner 9* focused on getting the spacecraft to orbit a planet (Mars in this case).

Verner Suomi became most interested in these planned explorations of neighboring planets. His interest has been concisely summarized by Larry Sromovsky (Figure 14.1), a leader of SSEC scientific teams that studied the atmospheres of the planets in our solar system from the mid-1970s to the present time:

> It was not too surprising that the meteorological focus of Vern Suomi and the center he and Bob Parent started should extend beyond the earth to

investigate these other interesting weather systems. Perhaps they might give us some insight into our own weather, and into our own climate history, and perhaps its future.[1]

Venus was the first focus of this expanding interest when Suomi was appointed a member of the *Mariner 10* imaging team in 1972. Venus was especially interesting to scientists because of its runaway greenhouse effect in a nearly pure CO_2 atmosphere. Further, the mission that began in November 1973 revealed detailed horizontal cloud structure that could be used to trace Venus's circulation, which surprisingly turned out to be super rotating and displayed an interesting polar vortex discovered by Suomi and his doctoral student Sanjay Limaye through use of "movie loops" from McIDAS (Limaye is now a senior scientist at SSEC).

Suomi's interaction with the imaging team is characteristic of his scientifically argumentative nature. Other members of the team—mostly geologists—were skeptical about the conclusion regarding circulation and stated that "just because it looks like a vortex, doesn't mean it is a vortex" (Research Sampler 1989/90). Suomi retorted: "Fine. [So] when you see those round basins all over Mercury, you can't call them craters." Later observations from a Venus orbiter confirmed the vortex circulation.

Suomi entrained Sromovsky and Hank Revercomb into the space-exploration ventures at SSEC in 1971 and 1973, respectively. Paul Menzel would join them in 1974. Interestingly, none of these physicists thought they were qualified to work in satellite meteorology. Sromovsky's story of joining SSEC is representative of stories from the other two:

I was finishing my dissertation on scattering theory of nuclear particles [in 1970] and jobs for physicists were few and far between at that time. Suomi presented a seminar at our Friday afternoon colloquium. I don't remember much about it but he ended by saying he needed people and he had funding. That motivated me to talk to him about a possible job. I asked for some of his recent papers to read so that I could better understand the nature of the work. After reading them, I came back and told Suomi that I didn't think I had the right background to contribute very much to his research efforts. He said 'don't be so narrow minded' and offered me a job anyway. That was a tipping point for me, setting me on a career path that I could not have imagined. It

1. This quote is from an unpublished SSEC technical report by Sromovsky titled "Planetary Science Research at SSEC, History and Current Focus."

wasn't too long before I got involved with the VISSR Atmospheric Sounder (VAS)—working on ways to fit the multi-band infrared radiometer into the existing geosynchronous imager to provide simultaneous imaging and temperature sounding. After that, Suomi asked Hank and me to start working on the net flux radiometer for the Pioneer Venus Mission. Paul continued to work on VAS. (L. Sromovsky 2015, personal communication)

The Pioneer Venus proposal from SSEC was written in 1973, and it included imaging and radiation measurements from a battery of instruments—but only that portion dealing with the net-flux radiometer was accepted. Suomi, Sromovsky, and Revercomb designed the net-flux radiometer. The radiometer was called a "lollipop," resembling a piece of flat hard candy stuck on the end of a four-inch stem. This flat piece, about the size and shape of a quarter, flipped back and forth every second so that measurements from both sides could be compared. If the measurements were not identical, then an average value was taken to remove the random error. Master machinist Bob Sutton put the radiometer together. As remembered by Revercomb, in the end it was like "a piece of jewelry." A group picture of the Pioneer Venus team is shown in Figure 14.2 where Suomi is holding the sensor.

As often happens when something entirely new is undertaken—as was frequently the case at SSEC—unanticipated problems develop. Getting the flip mechanism and sensor to function over the wide range of temperatures and pressures to be encountered during descent to the surface of Venus proved to be a challenge. Sromovsky and Revercomb carried out the testing of the external sensor hardware. The instrument was placed in a large cylindrical container into which 90 bars of CO_2 were pumped, while the interior was also being heated to 735 K, simulating the descent to the surface of Venus. These two youthful physicists worked for two years testing the instrument in a multitude of different configurations, helping the engineering team to eventually find a reliable design. The considerable time they spent in SSEC's dark basement led them to refer to themselves as "mushrooms" (H. Revercomb 2015, personal communication).

Other problems arose in the development of the instrument's electronics. At one point during an early stage of the project, the engineering team approached Suomi and delivered their consensus decision—the time allotted and cost for the required work was not going to be adequate. Irritated by this revelation, Suomi delivered an ultimatum with two options. He pulled out a legal pad and ripped two pages off the top. On the first sheet he wrote, "Dear NASA, we found that we are unable to complete the engineering task

Figure 14.1. From left: Henry Revercomb, Lawrence Sromovsky, and Suomi examine a photograph of Venus (courtesy of the Photographic Media Center, University of Wisconsin–Extension).

for this project, and we will return your money." On the second sheet he wrote, "Dear NASA, we're facing some unexpected obstacles in completing the engineering task on this project, and we find we will need added funds and a little more time to complete the project." He then left the room with the admonition for them to make a decision regarding the two options. As might be expected, they opted for the response on the second sheet of paper.

Suomi always had a flair for the dramatic, not only in motivating an action, but also in making a point. During one NASA review of the SSEC net-flux radiometer project, a panel member expressed doubts about the ability of the sensor to survive the 250-Gentry deceleration. Suomi responded by

Figure 14.2. Pioneer Venus net-flux radiometer instrument team. Back row, from left: Evan Richards, Gene Buchholz, Bob Herbsleb, Wanda Lerum, Jerry Sitzman, Hank Revercomb. Front row, from left: Doyle Ford, Tony Wendricks, Ralph Dedecker, Verner Suomi, Larry Sromovsky, Robert Sutton (courtesy of the Photographic Media Center, University of Wisconsin–Extension).

taking a sample sensor head and throwing it against the wall of the room with great force. The sensor was undamaged by this dramatic act and it made a great impression on the panel.

In 1978, the Pioneer Venus spacecraft reached the planet and the four heat-shielded probes were parachuted into Venus's atmosphere. The domed-shaped probes are shown during laboratory testing in Figure 14.3. These net-flux radiometers measured broadband combined solar and thermal net fluxes to determine the degree of radiative heating and cooling within the atmosphere as well as the atmospheric temperature during descent.

SSEC turned toward research on the outer planets (Figure 14.4) when Suomi became a member of the Voyager imaging team. The two Voyagers were launched in 1977 and reached Jupiter in 1979, Saturn and Titan in 1980 and 1981, Uranus in 1986, and Neptune in 1989. According to Sromovsky (see unpublished report cited previously), this could have been NASA's greatest scientific adventure. He retrospectively discussed seminal results from the Voyager missions:

Figure 14.3. Dome shaped probes that comprised the Pioneer Venus Multiprobe. Taken at Hughes Aircraft Company (courtesy of NASA Ames Research Center Image Library).

A few big surprises included numerous active volcanoes on Jupiter's satellite Io, a seasonal north-south asymmetry on Titan, the richly detailed structure of Saturn's rings, and the ring arcs and surprising Great Dark Spot on Neptune. We obtained detailed measures of the circulations of Jupiter, Saturn, and Neptune, but only a crude measure of Uranus' circulation due to its bland appearance and absence of discrete clouds that could be seen with the limited wavelength coverage of Voyager's crude vidicon camera system. . . . Textbooks had to be greatly expanded.

Figure 14.4. From left: Larry Sromovsky, Sanjay Limaye, and Bob Krauss examine interplanetary photographs at SSEC (photographer: Norman Lenburg; courtesy of UW–Madison Archives).

SSEC continues the quest for knowledge concerning the outer planets. Planetary scientists at SSEC have also expanded their approach by observing the outer planets with the Hubble Space Telescope and the ground-based telescopes at the Keck Observatory, making great improvements in our understanding of the circulation and composition of the atmosphere of Uranus. Indeed, Suomi's idea of expanding the vision of our own atmosphere's circulation by examining those of neighboring planets has paid off in a variety of ways. Not only was there stimulation in capturing the variance in circulation patterns of our neighbors, but the challenges and demands of constructing radiometers that could withstand the harsh environments of our neighboring planets' atmospheres has also expanded the capabilities of the engineers and technicians at SSEC. And beyond this, the presence of experimental physics and engineering alongside theory enlivened an already exciting place to work.

CHAPTER FIFTEEN

Satellite Data in Service to NWP

By necessity, the first operational NWP models developed in the mid-1950s were low-order systems with a limited number of grid points and a single level in the vertical. The dynamical constraint was the barotropic vorticity equation that governed the motion of long waves in the atmosphere (see Figure 6.3 in Charney et al. 1950). Nevertheless, these first operational models in Sweden and the United States stretched the available computer power to its limit (reviewed in Thompson 1983; Wiin-Nielsen 1991). Initialization of these deterministic models relied on observations from the radiosonde network—barely adequate over the European and North American continents and virtually absent over the oceans. Then with the World Weather Watch's lofty goal of developing extended-range forecasting with more-realistic models, there came the need for an observational system that could support the global models. This Herculean job and challenge fell on Suomi and his colleagues that served as members of GARP's Joint Organizing Committee. The observational system would need to rely on measurements made from artificial satellites that circled the globe.

Since the early 1960s, there have been two major thrusts to incorporate satellite data into NWP: 1) determination of satellite soundings (temperature retrievals) from observed infrared radiances in the 15-μm band of CO_2 and their input to the radiative transfer equations (Chandrasekhar 1950) and

2) determination of atmospheric motion vectors (AMVs) that approximate wind through cloud tracking. Suomi and his team at SSEC were front and center on the development of AMVs, but only tangentially connected to the issue of incorporating temperature soundings into data assimilation at operational NWP centers. Nevertheless, Suomi had given serious thought to combining temperature and moisture soundings with movies of weather systems through the VAS project. The emphasis of this project was analyzing weather pictorially and quantitatively from geostationary altitude on the scale of storms, the mesoscale. It did involve retrieval, and Bill Smith, a pioneer in the satellite sounding process, worked alongside Suomi after he and his NOAA/NESDIS team moved to SSEC in the late 1970s.

We start our discussion with the impressive work by the Australian meteorological community that innovatively used cloud pictures from *TIROS-1*, *Nimbus-1*, and *ATS-1*. Information in this imagery was used to make estimates of synoptic structure in the data-sparse Southern Hemisphere and eventually used in the forecast/cycling process at the Australian Bureau of Meteorology (BoM).

Use of Cloud Pictures

Some of the most impressive use of visible cloud imagery from the TIROS and Nimbus series of satellites in the 1960s and 1970s came from the BoM and the Australian Numerical Meteorological Research Centre (ANMRC) headed by Douglas Gauntlett. Important publications that describe the work of forecasters and research meteorologists are found in a conference presentation (Guymer 1969) and in the refereed literature (Troup and Streten 1972; Zillman and Price 1972; Seaman et al. 1977; Kelly 1978). Their approach is succinctly stated in the abstract of Kelly's paper: "A semi-objective procedure has been developed to modify mean sea level pressure and 1000–500-mb thickness using cloud vortex patterns obtained from satellite imagery. The method combines the previous work of Nagle and Hayden (1971) and Troup and Streten (1972), and is designed for operational use, particularly in the Southern Hemisphere . . . [where] a forecast/analysis cycle using this method with operational products of the National Meteorological Analysis Centre of the Australian Bureau of Meteorology [has been tested]." Bob Seaman, a research meteorologist at ANMRC, was a pioneer in objective analysis of weather over the Southern Hemisphere, and his creativeness was apparent in Seaman et al. (1977). In this paper, a computationally efficient variational analysis scheme was designed for operations, and it could accommodate the semi-objective estimates of thickness and sea level pressure along with in situ observations—a tour de force in those early days of data assimilation with limited computational power. An informative review of the work by the Australians is found in Bourke (2004).[1]

Satellite Soundings from Measured Radiances

We give some attention to the retrieval of temperature from satellites since it is central to the analysis/forecast cycling at operational NWP centers. Although the stages and steps in this saga are very interesting, it is a complex story that deserves its own historical study, and accordingly, we will be satisfied with an abbreviated review. We draw heavily on information in a comprehensive review by Smith (1985) in the *Handbook of Applied Meteorology* and an unpublished 30-year history on the impact of satellite observations, especially satellite soundings, on NWP by Eyre (2007).

Suomi and Parent were primarily interested in the radiative fluxes that determined the long-term heat balance in the Earth–atmosphere system, in essence, averages over long periods of time—months to seasons to years. In the case of satellite soundings and their impact on NWP, the radiances measured from satellites are still the primary observations, but the time scale of interest is hours to days to a week or slightly more. In the case of NWP, the radiative transfer equations are coupled with the equations of motion where Suomi and Parent had the luxury of neglecting this day-to-day coupling. However, Suomi and Vonder Haar coupled the long-term global balance and latitudinal imbalances with the general circulation transports in the ocean and atmosphere (chapter 13).

When we consider the process of creating temperature retrievals/soundings from measured radiances on a day-to-day basis, we depend on the theory that was presented to the atmospheric physics community by Lewis Kaplan in that age of fascination with space science and space travel (Kaplan 1959).[2] The principle rested on satellite-measured spectral radiances as forcing in the generalized radiative transfer equations (Hayden et al. 1979). Kaplan advocated the use of infrared radiances in the 15 μm

1. Bourke's study *History of NWP in Australia – 1970 to the present* is available from National Library of Australia: https://catalogue.nla.gov.au/Record/3354061.

2. The idea for calculating atmospheric temperature soundings from satellite-observed radiances and radiation theory was presented by J. I. F. King (1958) a year before Kaplan, but it received less attention from the atmospheric science community.

band of carbon dioxide (CO_2) since the rich structure in this band supplies information on the temperature profile throughout the atmosphere in clear-sky conditions. Although appealing in theory, limitations and problematical aspects came with implementation. Among the problems were the following: 1) cloud opacity with respect to infrared radiation made retrieval in partly cloudy sky difficult and impossible in overcast conditions; 2) the absence of a unique solution to the radiative transfer equations for a finite set of measured radiances; 3) an optimal solution based on regularization required a guess profile (a nearby radiosonde-measured profile if available), and consequently, the optimal solution was dependent on the guess, and 4) to obtain high resolution in the vertical, radiances from a great number of frequencies (wavelengths) are required, and in practice only 6–8 frequencies were used and they yielded a vertical resolution of ~3 km.

Overcoming most of these limitations has been nearly a half-century undertaking (early 1970s to the late 1990s). When satellite soundings came on the scene in the early 1970s, the mechanics of creating initial conditions for deterministic NWP was geared to handling radiosonde (raob) observations. Efficient and empirical data assimilation schemes developed by Bergthórsson and Döös (1955) in Sweden and Cressman (1959) in the United States used the 12-h forecast as background and interpolated increments (difference between the raob observation and the forecast) to the grid points. It was natural to view the satellite soundings in this light, that is, make them appear as raobs. Of course, the strength of the satellite sounding is global coverage (from a polar-orbiting viewpoint initially but from a geostationary viewpoint later), but its weakness was poor vertical resolution—just the opposite of the raob observation.

The origin of this idea of making the satellite sounding appear as a raob is remembered by Ron McPherson, head of the assimilation branch at National Meteorological Center (NMC) in the 1970s through the 1980s:

> There were many problems with these observations [satellite soundings]. They were often contaminated by clouds . . . and they used a relatively small number of frequencies so that they exhibited too low resolution in the vertical. . . . In May 1969 I was called into Fred Shuman's [Head of the Development Division at NMC] office for a meeting between Fred and Dr. William Smith of NESS . . . Bill Smith was a leading scientist on design of satellite radiance—measuring instruments—an excellent scientist, passionate about their use and given to overselling their value. He was there to persuade Fred that NMC should use these data . . . after listening to Smith's presentation, he

said, "All right, Bill, if you can make these data look like a radiosonde, we'll try to use them." (R. McPherson 2016, personal communication)

After incorporating these "radiosonde look-alike" satellite soundings into assimilation at NMC, the Data Assimilation Branch ran a series of tests to assess the value of these retrieved soundings—called Data Systems Tests (DSTs). In these DSTs, parallel runs of the operational prediction system were executed, one that included satellite data and the other without these data. The results were mixed in a three-way competition between NMC, GFDL, and the NASA Goddard Space Flight Center (GSFC). The GSFC and GRDL systems showed positive impact while the NMC was slightly negative.

By the mid- to late 1970s, the Bergthórsson–Döös and Cressman empirical assimilation schemes at the operational centers gave way to the optimal interpolation (OI) (Gandin 1965) or so-called statistical interpolation (SI) schemes (Rutherford 1972). Although these schemes were optimal (determination of weights on the increments that minimized the ensemble error of the estimate) instead of the empirically determined weight, they still treated the satellite soundings as point observations (look-alike raobs) with error characteristics disconnected from the intrinsic error in observed radiances. The difference between temperatures of satellite soundings and collocated raobs was on the order of 2°–3°C and, as might be expected, the summary of seven observations systems experiments (OSEs) that included or withheld satellite soundings indicated that insertion of satellite soundings into the assimilation system led to small improvements in the forecasts (Ohring 1979).

At the European Centre for Medium-Range Weather Forecasts (ECMWF) seminar "Data Assimilation and Observing System Experiments with Particular Emphasis on FGGE" in 1984, a consensus opinion stated that satellite soundings were playing an important role in analyzing large-scale weather systems (especially in the Southern Hemisphere), despite the fact that the accuracy of the soundings was relatively poor. But the real crisis came in the late 1980s when the prediction models improved considerably. As stated by Eyre (2007): "As a consequence of the [greatly improved] short-range forecasts . . . the potential for damage through erroneous observations or observations assimilated in an inappropriate way, was much greater." In this situation, "the satellite data started to have either an average negative impact or at least an impact that was very variable from case to case" (Eyre 2007). Suspicion began to surround the OI and SI assimilation systems in this era of outstanding model forecasts. The scheme is linear and the computational demands are great, with the need to invert extremely large matrices that measure the error covariance of forecast and observations. The variational scheme overcomes both of these limitations, but iteration toward optimality is computationally demanding (Eyre 1997; Lewis et al. 2006). Work on the variational assimilation scheme started in the early 1990s (Eyre et al. 1993), and with evidence of problems with OI (Kelly 1997), a decision was made to switch from OI to variational assimilation methods. The switch took place at the National Centers for Environmental Prediction (NCEP) in 1995 (discussed in Derber and Wu 1997) and at ECMWF in 1997 (discussed in Anderssen et al. 1994). With the variational system, the model counterparts to observed radiances are accurate (without need of linearization) and errors in the radiances are intrinsically accounted. In the introduction of Simmons and Hollingsworth (2002), there is a valued presentation of verification statistics for the NCEP and ECMWF variational assimilation systems (Figure 1 in their paper). Results from both centers show rather impressive reduction in 3–5-day forecast errors for geopotential after installation of the new assimilation systems. Most impressive is the dramatic reduction in ECMWF's 3–5-day forecast error in the Southern Hemisphere's 500-hPa geopotential.

Satellite-Derived Winds

Suomi began to think about cloud-drift winds as input to NWP in the mid-1960s, when the power of the spin-scan camera had been demonstrated. Tracking a cloud and noting its displacement from a motion picture was easy enough by eye, but a method that accurately measured displacement and direction over a given interval of time was not so obvious, nor was it easy to determine the height of the cloud. In the midst of thinking about this use of satellite imagery, he was surprised to find that his colleague, Professor Theodore "Ted" Fujita, had already used ATS imagery at 20-min intervals to capture cloud motion in association with tornadic storms. Suomi's astonishment is captured in the following quote from his oral history interview:

> Ted Fujita had a wonderful photo lab down there [at University of Chicago], and he is a fabulous photographer, knows how to combine things and put grids on and all that. He made a movie; he simply beat the pants off me with my own satellite [spin-scan camera]! I didn't like that. (Suomi et al. 1994)

Satellite-Derived Winds Development at SSEC: An Historical Perspective

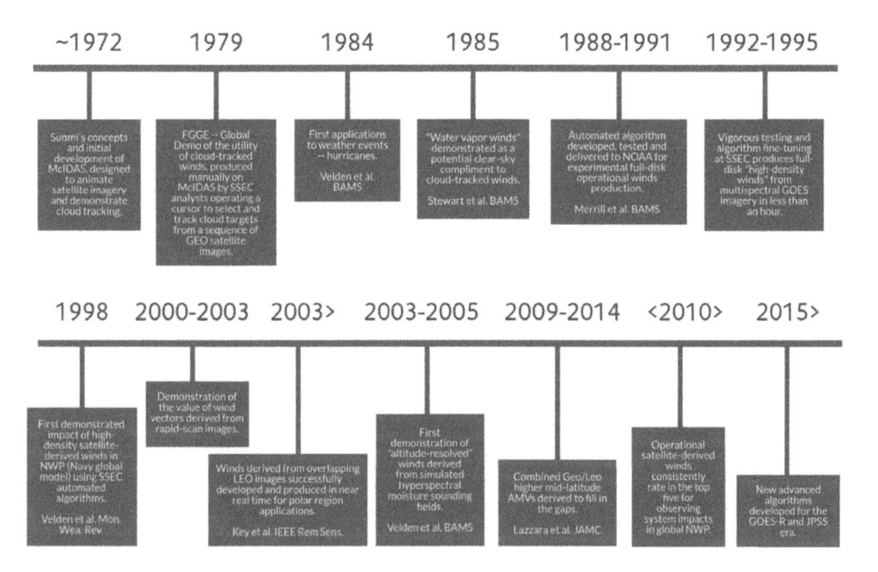

Figure 15.1. Timeline of AMVs development at SSEC with major milestones.

Indeed, Fujita spent a considerable amount of time extracting remotely sensed data from artificial satellites to get a more complete analysis of mid-latitude mesoscale storm systems and tropical storms. His large body of work in this domain is masterfully captured in Menzel's (2001) review article. The movie that is referred to in the quotation above "was widely shown and demonstrated to the meteorological community" (Menzel 2001). For those of us who witnessed Fujita's talk on this motion picture produced by consulting meteorologist Walter Bohan (Fujita and Bohan 1967), the presentation was filled with excitement and humor, but it convincingly caught the power of weather movies from space.[3] Not to be foiled, Suomi started to think about using calculations of cloud-drift winds based on digital images rather than the analog images used by Fujita. His thoughts along this line eventually led to McIDAS. A schematic that follows the chronological development of AMVs is found in Figure 15.1.

It should also be mentioned that much of the work with AMVs was coordinated through the combined efforts of the Cooperative Institute for

3. The author (J. L.) attended Fujita's seminar on this topic at the National Severe Storms Laboratory while a graduate student at the University of Oklahoma.

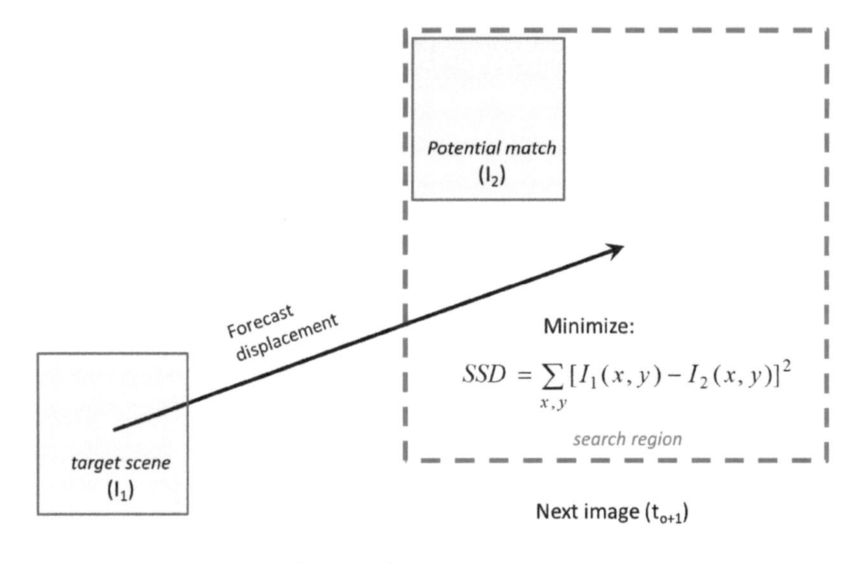

Figure 15.2. Schematic showing the basic concepts associated with the feature-tracking algorithm used at NOAA/NESDIS. Targets are selected from the middle image of a three-image loop and tracked forward and backward in time using a correlation method. The two displacements are averaged to produce a final motion estimate. Only the forward vector is shown in the figure.

Meteorological Satellite Systems (CIMSS) and SSEC.[4] In the early 1970s (around 1972, as indicated in the box to the far upper left in Figure 15.1), Suomi instructed SSEC programmers and "budding satellite meteorologists" [Gary Wade, Tony Schreiner, Todd Stewart, and Chris Velden, all MS (master of science) students under Suomi] to work together and develop software that could co-register and navigate cloud element tracks on the ATS imagery. At this stage, tracking was manual, where a "joystick" in the hand of the meteorologist defined and registered the locations of the cloud element. While crude, the concept of AMVs was born (Figure 15.2). The power of the concept was proved when AMVs supplemented in situ wind observations in the First GARP Global Experiment (FGGE) datasets that were being used to analyze tropospheric winds over the tropical oceans.

4. CIMSS was created through a Memorandum of Understanding between UW–Madison and NOAA and later, NASA. As an institute within SSEC, this memorandum gave Bill Smith's NOAA group a secure place within the UW–Madison system of higher education.

Figure 15.3. Photo from September 1985 at the National Hurricane Center, Coral Gables, FL. Hurricane specialist Max Mayfield (seated) is operating the newly installed McIDAS system, demonstrating satellite imagery of Hurricane Elena to Governor Bob Graham (standing with hand to his chin) and the governor's wife, Adele. Others who are standing and viewing the imagery are (l. to r.): Robert Sheets (NHC Deputy Director), NHC Director Neil Frank, unknown, Graham, and SSEC scientist Chris Velden.

Following this success with FGGE data in the early 1980s, the emphasis shifted to manual production of AMVs over targeted areas in near-real time. Demonstrations of this process took place at operational centers like the NHC and at research laboratories like the National Severe Storms Laboratory (NSSL) during their springtime field programs. One such demonstration took place at NHC in early September 1985, when Hurricane Elena was bearing down on the southeastern United States. NHC forecaster Max Mayfield is shown generating satellite products associated with Elena in Figure 15.3. The governor of Florida, Bob Graham, is shown paying rapt attention to the process along with NHC forecaster Bob Sheets, Neil Frank (NHC director), and SSEC's Chris Velden, the McIDAS instructor. Velden trained most of the NHC forecasters in the use of McIDAS (including the generation of AMVs) after a prototype system was installed at the center in 1984. Further, NHC requested that Velden be "on call" at NHC to provide expertise in the use of McIDAS when hurricanes threatened the United States. SSEC and CIMSS honored the request, and Velden has spent considerable time at NHC from the mid-1980s to the present day.

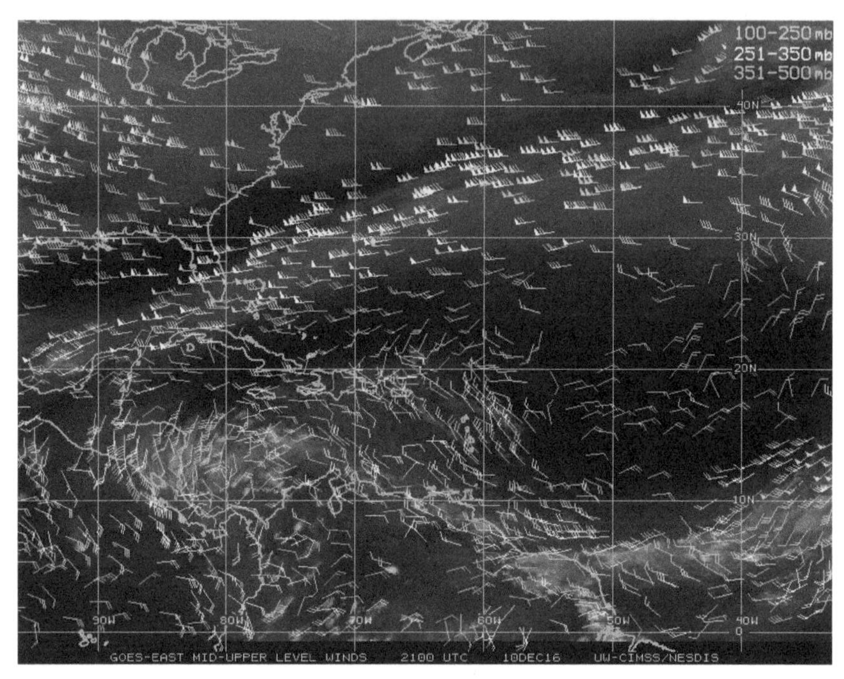

Figure 15.4. Plot of upper-level AMVs on December 10, 2016 from GOES-East over a water vapor (WV) image, color coded by height: 100–250 hPa (cyan), 251–350 hPa (yellow), 351–500 hPa (green). Vectors are plotted in knots.

Other operational prediction centers such as the National Severe Storms Forecast Center [NSSFC in Kansas City, Missouri; now the Storm Prediction Center (SPC) in Norman, Oklahoma] benefitted from the demonstrations and now have forecasters who are trained in the operation of McIDAS and generation of AMVs. Clearly, Suomi's McIDAS system was an integral part of the NHC and NSSFC modernization and provided the prototype for the National Weather Service's Advanced Weather Interactive Processing System (AWIPS).

Automation of the manually intense process of creating satellite-derived winds was the next major hurdle to overcome. "When Suomi hovered behind a human operator in front of the McIDAS terminal, it would take 1–2 hours to complete a set of vectors over the USA, for example" (C. Velden 2016, personal communication). The breakthrough came from the innovative work of SSEC research meteorologist Bob Merrill—a programmer extraordinaire. He coded an objective scheme on McIDAS that would create AMVs over a sizable region in minutes rather than hours—an order of magnitude reduction in time.

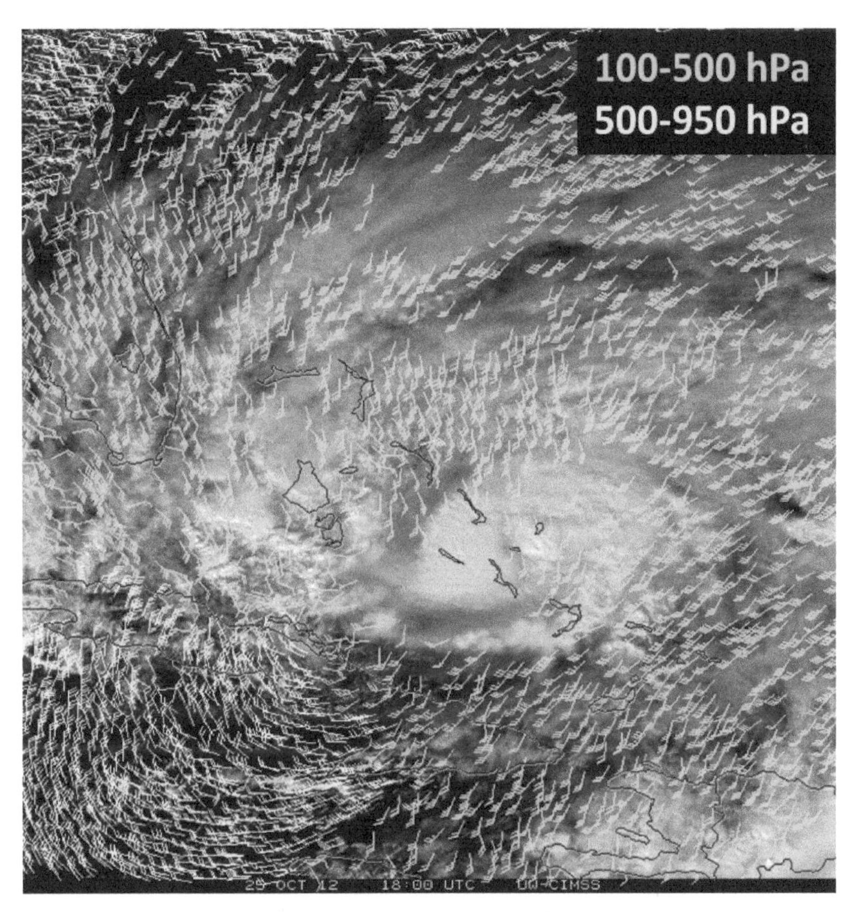

Figure 15.5. Example of high-resolution VIS/IR (visible/infrared) AMVs derived from GOES 3 - 5 minute super-rapid-scan imagery during Hurricane Sandy in 2012.

With automation came the real-time production of fields such as those shown in Figure 15.4. The rapid production of these fields made them attractive for operations. At SSEC, the first operational model chosen for testing was SANBAR, a steering current model developed by MIT professor Fred Sanders and his doctoral student Bob Burpee (Sanders and Burpee 1968).[5] SANBAR (short for "Sanders Barotropic"), was used operationally to track hurricanes at the Hurricane Research Division (HRD), a NOAA research laboratory collocated with NHC. AMV winds were used as initial

5. The origin of this idea by two graduate students at the University of Tokyo in the mid-1950s and its use in operations at the Japan Meteorological Agency (JMA) is historically reviewed in Lewis and Lakshmivarahan (2008).

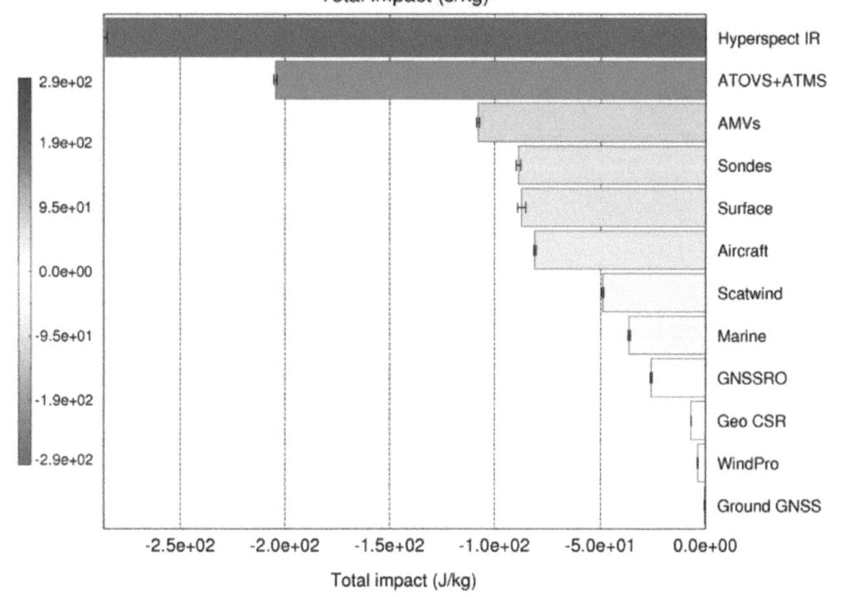

Figure 15.6. Example from the Met Office (United Kingdom) showing forecast sensitivity to observations impacts (FSOI) from an operational model suite using an adjoint-based method for estimating observation impacts. The measure is the impact on 24-hour forecast error in terms of a global, total (moist) energy norm calculated from the surface up to 150 hPa. Categories of observations are ranked by their impact importance. Looking back on the years since atmospheric motion was introduced into NWP, we find that the forecast skill realized for 3 days out is now being realized for 7 days out.

conditions for SANBAR, and the track forecasting exhibited modest positive impact (Lewis et al. 1987). The results attracted attention from research meteorologists at operational NWP centers, and AMVs were given serious consideration as input to the more complicated models. The U.S. Navy's global model was one of the first to go operational with the AMVs produced by CIMSS–SSEC and realized positive impact (Goerss et al. 1998; Velden et al. 1998). These wind fields were attracting attention in the international community as well. Working with ECMWF's variational assimilation system, Kelly (1997) showed that insertion of AMVs had a positive impact on global forecasts, especially in the Northern Hemisphere and tropics.

Since the turn of the century, the focus has shifted from development to new innovations and enhancements of the AMV product. Recent advances in satellite sensors with more stable image registration and scanning strategies allow higher temporal refresh rates, and this has led to better cloud targeting

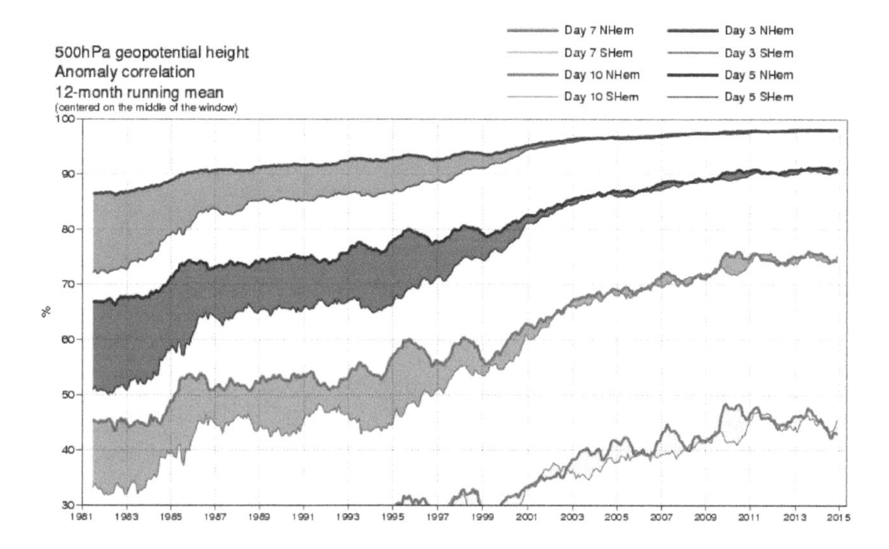

Figure 15.7. Annual running mean of anomaly correlations (in percent) of the 500-hPa height forecasts for 3-, 5-, 7-, and 10-day forecasts for the extratropical Northern and Southern Hemispheres for the ECWMF operational forecasts from January 1980 to May 2013. The shading shows differences in scores between the two hemispheres (from Dee et al. 2014).

and tracing. Coupled with improvements in tracking and height assignment methodologies, along with more elaborate quality-control procedures, AMV observation errors have been considerably reduced.

Novel applications include the derivation of AMVs from rapid-scan geostationary (GEO) imagery for smaller-scale applications (Velden et al. 2005; Figure 15.5), over polar regions from low-Earth-orbiting (LEO) satellites (Key et al. 2003), and higher midlatitude AMVs from combined LEO–GEO (Lazzara et al. 2014).

All of the research and development advancements have paid off: AMVs consistently rank in the top five of all tropospheric observation types in most global NWP center impact studies (Figure 15.6). This is testament to Suomi's wisdom that measuring the Earth's winds was vital to weather forecasting, and to his original vision that this could be accomplished from satellites in space.

Figure 15.7 shows the evolution in time of the anomaly correlation of the 500-hPa height forecasts at the ECMWF (this is a standard parameter for the ECMWF to measure impact and improvement). The increase in forecast skill is shown for different forecast ranges. Improvements are clearly discernable, and they are the result of improvements in measurements providing better and/or new data, data assimilation and modeling, and computing resources. Also evident is the gradual equalizing of the forecast skill for the Southern Hemisphere (largely relying on satellite data only) and the Northern Hemisphere (where satellite data complements a robust ground based observing system).

V

Suomi's Uniqueness

CHAPTER SIXTEEN

Curtain Call: Last Project and Last Days

Verner Suomi gave up the reins of SSEC directorship in 1989 at the age of 74. The decision was dictated in part by failing health—a history of heart ailment. Francis Bretherton, British geophysical fluid dynamicist and former president of the University Corporation of Atmospheric Research (UCAR), took over the helm at SSEC. Suomi was now emeriti on two counts: Harry Wexler Distinguished Chair Professor at UW–Madison and Director of SSEC. He continued to comb the halls of SSEC and strike up conversations with the workforce. In fact, he seemed more relaxed and put significant effort into teaching the general education atmospheric science course, where he stimulated undergraduates to think about careers in science. It was not unusual for him to start the course with the following whimsical but true statement: "Don't be afraid of science and don't be intimidated by it; use your intuition, that's what I did with only a single course in college physics." Indeed, when Suomi told physics chair professor Ragnar Rollefson this story, Rollefson replied, "you've surely gotten a lot of mileage out of freshman physics" (Gregory 1998).

Suomi's cheerful state during this period is captured in the photo taken on May 18, 1990 (Figure 16.1). At this time, he was satisfied that the first phase of GEWEX (Global Energy and Water Exchanges)—the buildup phase—had just begun. Suomi had been disappointed in NASA's drift away from the

Figure 16.1. Vern Suomi in his office at SSEC working at his computer (1990) (courtesy of UW–Madison).

reality of weather phenomenon toward the abstraction of "global change science," and in collaboration with Pierre Morel and Lennart Bengtsson, the concept of GEWEX had been proposed (P. Morel 2006, personal communication). The project is dedicated to understanding Earth's water cycle and the energy fluxes in the Earth–atmosphere system. In essence, Suomi and his colleagues were following Rossby, whose last-planned research thrust was a move from Stockholm to the Middle East to concentrate on the world's water issues (letter from Harriet Rossby Woodcock, 1992, Figure 16.2). Rossby died in 1957 and the plan was never executed.

The meeting in Suomi's office on May 18, 1990 was designed to obtain oral history on Rossby. And indeed, the interview commenced with the following question for Suomi: When the name Carl-Gustaf Rossby is mentioned, what is your first thought? Suomi's reply: "He could talk you out of your shirt and you thought you were getting a good deal!" What could be more apropos? Suomi was describing a feature of his own style of mentorship: entrain the student and co-opt the student into your current interest. As recalled by Menzel (2009, personal communication),

> As was well known, Suomi had youthful associates who were brought into his fold. He'd invite us to lunch with visiting dignitaries at the [nearby restaurant]

Dr. John M. Lewis
National Severe Storms Lab.
1313 Halley Circle
Norman, OK 73069

Pohai Nani
45-090 Namoku St.
Kaneohe, Hi 96744-5320
March 1, 1991

Dear Dr. Lewis,

Thank you for your letter and the information about your proposal which seems to me to be a fine undertaking. I am sorry to be so slow answering your request.

Maybe too slow! Anything I can add may just be affirm what you already know. You seem to have tapped all the sources and probably people still available.

Carl made a point of not staying in one place longer than ten years and not that long in most. I think the Middle East was the next place in mind. This was almost nomadic as far the family was concerned and never easy shifting places

Figure 16.2. Letter from Harriet Rossby Woodcock, 1992.

> Brat und Brau (e.g., NASA and NOAA directors and scientists from around the world), encourage us to contribute to the conversations over lunch, and later that week we'd be working away on that latest project.

Suomi did indeed deliver important information about Rossby during this interview, but as typical of Suomi, he turned the conversation from history to current events. And what was the centerpiece of current events? It was his

Figure 16.3. Suomi describing his Frisbee flux meter in the parking lot adjoining SSEC (1990).

interest in calculating fluxes from the ocean surface with a strange contraption consisting of a small, flat circuit board surrounded by a set of concentric Styrofoam circles (see Figure 16.3). He called it his "Frisbee flux meter." As he said, "Just think of it, you're on the deck of a ship and you twirl this thing out over the sea just like you'd toss a Frisbee and it records the heat flux at the air–ocean interface and transmits the data to a satellite" (V. Suomi 1990, personal communication). The instrument was officially called the Skin-Layer Ocean Heat Flux Instrument (SOHFI).

Suomi got the idea for SOHFI in the mid-1980s. As typical of Suomi, his initial experiment with the idea was low budget and involved testing in his backyard with the grandkids' plastic swimming pool and eventually in the house. As recalled by Fred Best, "Suomi's early development and testing was conducted at home. One day he confessed to me that his wife Paula [Figure 16.4] really wanted her bathtub back." He then worked with Bob Sutton down in SSEC's machine shop to design the instrument.

At the SSEC's 50-year celebration in late 2015, Best made a sterling presentation that reviewed work on the SOHFI. Key statements in his presentation follow:

- Just below the air–sea boundary, there is a region in the skin of the water called the conduction zone, where the temperature changes

Figure 16.4. Paula Suomi.

linearly with depth. The trick is to place the heat flux sensor in this conduction zone. The heat flux sensor used is called a thermopile, and it directly measures the difference in temperature between its top and bottom surface, which, because we are in the linear conduction zone, is proportional to heat flux.[1]

- The entire deployed instrument includes the tethered heat flux float assembly attached to a spherical buoy that contained the support electronics, including a transmitter for satellite communications system. A drogue chute was attached to the buoy to provide better coupling to ocean currents, reducing the influence of the wind on trajectory of the assembly. The heat flux sensors were supported by a fiberglass open-weave mesh tensioned by a float ring. Buoyancy holds the sensor just below the surface of the water and within the conduction zone (see Figure 16.5).

Larry Sromovsky became involved with SOHFI in the early 1990s. As he recalled,

1. This result follows from mixed layer theory: Kraus and Turner (1967) for the ocean and Lilly (1968) for the atmosphere.

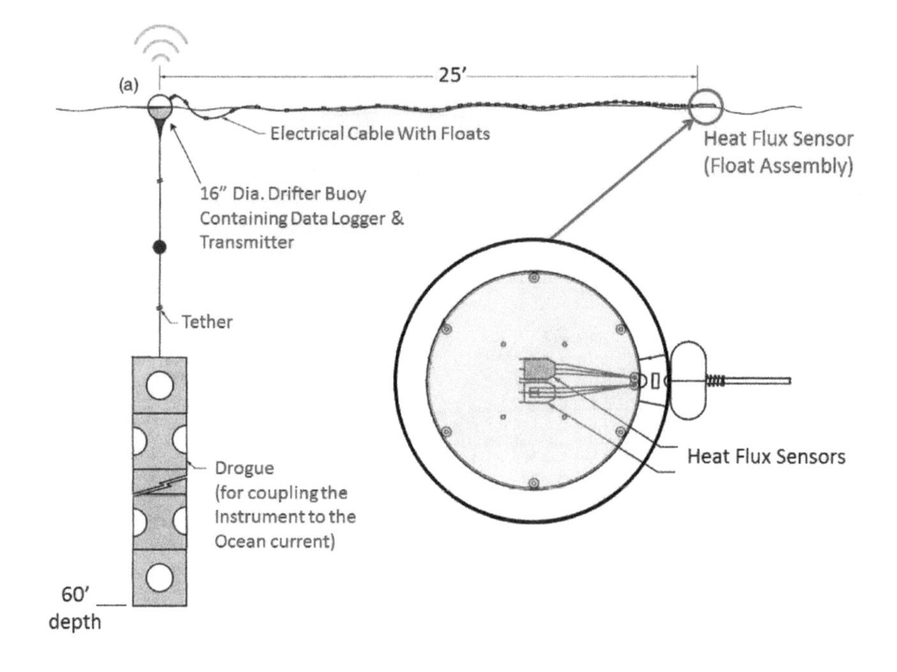

Figure 16.5. Schematic of the entire deployed instrument that included the tethered measurement sensor float attached to a spherical buoy that contained the support electronics, including a transmitter for a satellite communications system (courtesy of Fred Best, SSEC, UW–Madison).

> I was asked by Suomi (maybe by Fox and Bretherton too) to 'weigh in' on the project. I initially thought it was kind of an unworkable idea and it was hard for me to believe that it could actually be measuring anything accurately. I decided to look at some of the data collected from the instrument—data collected over some northern Wisconsin lake [Sparkling Lake]. After looking at the observations and comparing them with observations from nearby airfields, I came to believe that it was measuring something reasonable. Anyway, I kind of gave my blessing to the idea after coming in as a skeptic . . . Suomi got a big kick out of that. (L. Sromovsky 2015, personal communication)

Suomi became energized after Sromovsky's analysis of the Sparkling Lake data and wrote proposals that secured SSEC bridge funding for the project. But progress on SOHFI was slow, it was unfunded from outside agencies, and it was burning money from overhead at a good clip. The team of engineers and technicians working on the project was sizable and included Best along with Evan Richards, Jerry Sitzman, Bob Herbsleb, and Gene Buchholtz (see

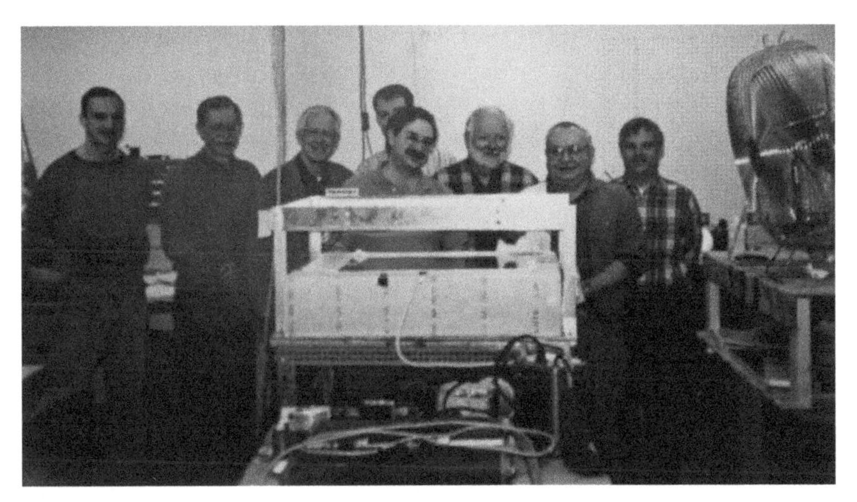

Figure 16.6. The SOHFI team (from left): Chris Cisko, Larry Sromovsky, Verner Suomi, Mark Mulligan, Fred Best, Jerry Sitzman, Gene Buchholz, Evan Richards. Team members not shown: John Anderson, Nick Ciganovich, Ron Koch, Hank Revercomb, Paul Wisniewski (courtesy of SSEC, UW–Madison).

Figure 16.6). Bretherton was concerned about the budget and Suomi was concerned about the lack of progress on making the instrument suitable for deployment at sea.

In early 1995, Suomi's health took a serious downturn and he was hospitalized in July 1995. As recalled by Sromovsky,

> Vern was dying of congestive heart failure but he kept asking me to bring in the latest instrument design to his room in the hospital. He never took a "play off"—like in football when some players relax on given plays whereas others never relax, that was Suomi [Suomi died on 30 July 1995].

Sromovsky made the first formal presentation on the instrument at the American Meteorological Society's Conference on Air–Sea Interaction in February 1996 [Suomi was posthumously listed as first author (Suomi et al. 1996)]. Deployment of SOHFI took place in the Atlantic Ocean during 1995–1996 and in the Pacific Ocean in 1997 (Sromovsky et al. 1999a,b). The following statement in the abstract of Sromovsky et al. (1999b) captures the state of the system in 1997: "The Gulf Stream and Greenland Sea deployments pointed out the need for design modifications to improve resistance to seabird attacks. Better estimates of performance and limitations of this device require extended intercomparison tests under field conditions."

Graduate student James Boyle, a coauthor on the papers mentioned in the previous paragraph, made further tests with the SOHFI on Lake Mendota and received his doctoral degree for the research. The project has lain idle since that time. Beyond any doubt, had Suomi lived into the late 1990s when these field tests were taking place, he would have come up with a "gadget" to discourage the birds, and he would have worked as tirelessly as possible to keep the project alive. In fact, there is evidence that Suomi was using all of his cunning to acquire financial support for the SOHFI project shortly before his death. In Fred Best's presentation mentioned above, he ended by quoting from the last paragraph of a proposal Suomi had submitted to NOAA. Referring to himself in the third-person grammatical style, Suomi wrote: "The PI is approaching his 78th birthday. He can't last forever . . . [however], as old as he is, he feels like a 16 year old with a new toy. Please let him enjoy."

CHAPTER SEVENTEEN

Epilogue

During his lifetime, Verner Suomi became legendary in both science and engineering, and in the years since his passing, the legend continues to grow. His nature could be deemed mutable and changeable—inconsistency was not a stranger—and we have labored to rescue the real, wonderfully complex nature of the man.

Suomi's Character

While Suomi presented himself as one of the working class, and indeed he came from the blue-collar Mesabi Range, he had ambition and intellectual drive only matched by his innovation and hard work in a high-level scientific sphere. His character was defined by both austerity and gregariousness. He could retreat into the quiet of a research idea, polished and better defined through communication with his brilliant engineering colleague Robert Parent or his "filters," but then balanced by a gregariousness where he drew others to him and co-opted them into his interest.

Suomi found his calling under Rossby at the University of Chicago in the 1940s. It can convincingly be argued that none of Rossby's protégés inherited so many of his characteristics as did Vern Suomi: an ability to identify challenging and critically important geophysical problems well before others, an approach to teaching that inspired yet was short on didactics and systematics, that aforementioned ability to co-opt a wide range of scientists into his vision, and that elusive quality to convince academic leaders and government bureaucrats to the worthiness of his scientific ventures. He had a foot in two worlds, the world of a dedicated researcher and visionary in academia and the world of the plain-speaking man, and this gave him advantage.

He could mix humor with science and excelled as master of the media soundbite. When asked about his entrée into satellite-based estimates of the Earth's radiation budget, he replied, "Oh, that was simple, all it took was two ping-pong balls, one painted black and the other white and hang them from a satellite." Yes, an overly abbreviated explanation, but one free of long-winded academic prelude—his style. And late in his career, when approached by a journalist with the question, "What is it you do?" His answer was "I take pictures of the Earth."

Preparedness

A major component of Suomi's approach to science and engineering was a nearly insatiable desire to know how things worked. It was seen in the disassembling of his sister's watch that she received as a Christmas gift, and it was evident during his last semester at Eveleth High School, when he single-handedly repaired an old X-5 airplane engine that had not worked for five years. Even more impressively, in his last days he designed and worked on a heat flux meter that could withstand the harsh environment of the open sea. As this last activity proved, these interests and traits remained with him throughout his career.

In Suomi's case, where we consider the skill as taking apart mechanical and electrical machinery, understanding its operation, and trying to put it back together, it is likely that Suomi had passed Malcolm Gladwell's 10,000-hour rule well ahead of the time when he assumed his professorship at the University of Wisconsin–Madison in 1948. At that time, Suomi embodied the normal academic strengths of a science or engineering professor at a research university, but he also possessed the skills of a master machinist or electrical technician. And these skills and interests coincided with the world's fascination with space science and space travel and the industry being built to support it.

Then, in 1953 during his doctoral exam at the University of Chicago on the subject of heat balance over a cornfield, an experiment he conducted over the Marsh Farms at UW–Madison, doctoral committee member Professor

Herbert Riehl asked the profound and stimulating question (in essence): Could you consider conducting this experiment over the Earth and its atmosphere? Of course he could, and the wheels of progress in this direction commenced to move immediately. The universal interest in space science that was now coupled with Suomi's new interest brought about that mystical process where developments in human affairs, cultural changes, and their associated needs coincide with the natural talents of an individual. In this case, it was coincidence between space science and the talents of Verner Suomi. He would design new and innovative instruments that could fly aboard artificial satellites and make measurements over the Earth and its atmosphere, and by so doing, he would strive to answer questions about its climate and weather.

Problem Finding and Problem Solving

The problems that Suomi chose to study were typically 1) majestic in scope—the heat budget of Earth and its atmosphere; 2) investigated from different angles of attack—similarity and differences between Earth's heat budget and those of neighboring planets; and 3) differing instrument designs to investigate the heat budgets.

The single problem, or we should say, a problem with variations on a single theme, took place over decades. In the next section of this chapter, the evolution of instruments takes center stage, and in this example we have just highlighted, the instruments for studying the heat budget of Earth, Jupiter, and Venus were related but different based on the characteristics of the atmospheres. Suomi's tendency to keep examining a challenging problem had similarity to the way mathematician Carl Gauss chose his mathematical problems: "[he] preferred to polish one masterpiece several times rather than publish the broad outlines of many as he might have done" (Bell 1937).

Suomi knew how to find "the problem," as sociologist Harriet Zuckerman stated in her study of Nobel Laureates in physics (Zuckerman 1977). She found that it was not knowledge or skills that protégés acquired from their masters so much as a "style of thinking." It was "problem finding" as much as problem solving. Identification of important and challenging problems came to Suomi before he took an active part in the Joint Organizing Committee of GARP, but there is little doubt that this activity was stimulating. If we look downstream from the early 1960s when Suomi became a key member of the JOC, its influence on problem finding is evident. The general circulation models of the 1960s, research models that were the prototypes of future extended-range NWP models, needed temperatures and winds on a global scale to define their initial condition. Of course, Suomi was in the position of planning for these global observations from artificial satellites. Based on these issues, Suomi began to think about the problem of getting wind estimates by tracking clouds. Over the next 15 years (mid-1960s to early 1980s), Suomi led SSEC in first developing the spin-scan camera that provided moving pictures from geostationary altitude—tracking clouds to get wind estimates—and then developing the data-processing system McIDAS that could handle the stream of data and produce the winds within the time constraints of operational NWP. This ultimately stimulated the work on VAS, a project that lasted another 15 years (from the early 1970s to the mid-1980s).

The problems he found and the problems he solved not only contributed to the success of GARP and extended-range forecasting, but they benefitted society. When these problems and their solutions are collectively examined, it is not surprising that Suomi was awarded the National Medal of Science in 1977.

Suomi Instruments and Inventions

For those who worked with and around Suomi, it was clear that he was happiest and in the best disposition when he was working on an instrument, talking to an engineer or machinist about instrument design, or inventing an instrument. His skills with instruments are remembered by Henry Revercomb (2014, personal communication): "He was an intuitive instrument builder, knowing how to cleverly take advantage of new technologies and use old technologies in new ways. He always worked from the principle 'keep it simple.'" We have discussed quite a few of Suomi's instruments and Suomi–Parent instruments in earlier chapters, but it is interesting to follow the evolution of these instruments. Revercomb has categorically subdivided Suomi's instruments and inventions and followed their evolution, as displayed in the table.

Let us add some detail to the information in the table. First, the *Explorer VII* bolometers were the product of collaboration between Suomi and Parent, as were the spin-scan camera and the flat-plate radiometers.

Column 1:

- The net-flux radiometers looked both upward and downward and were used in IGY follow-up experiments at the South Pole that involved Suomi's doctoral student Peter Kuhn.
- The probes for the Pioneer Venus Mission were built at SSEC, and they entered the Venus atmosphere on December 9, 1978. The calculation of net

Suomi's Instruments and Evolution

Net-Flux Radiometers Atmospheric heating and cooling		**Black & White Radiation Sensors** Earth's radiation budget		**Spin-Scan Camera** Measure atmospheric motion		**Stationary Sounder** Measure rapid changes in vertical profiles	
1953	Suomi's dissertation	1953	Inspired by Suomi's dissertation	1966	Camera development	1971	VISSR
1959–63	Suomi–Kuhn South Pole	1959	*Explorer VII* bolometers	1966	Flies on *ATS-1*	1980–94	VAS added to imager
1978	Pioneer Venus Multiprobe Mission	1966–72	Flat-Plate Radiometers ESSA, ITOS, NOAA satellites	1973	McIDAS (data processing)	1994–present	Sounder and imager separate
1994	Galileo Mission Jupiter	2016	Climate benchmark (CLARREO)	1980–present	AMVs		
		2016	Absolute Radiance Interferometer (ARI)				

flux depended on cloud, chemical composition, and a temperature distribution, and the calculations required a complex radiative transfer model.
- The radiometer entered the Jupiter atmosphere on December 7, 1995. These were the first in situ measurements of radiative fluxes in the atmosphere of Jupiter.

Column 2:
- Many satellite flights carried the Suomi–Parent flat-plate radiometers: *ESSA-3*, *-5*, *-7*, *-9* (Environmental Science Services Administration; 1966–1969); *ITOS-1* (Improved TIROS Operational Satellite; 1970); and *NOAA-1* (1970).
- The *Explorer VII* Radiation Balance principle carried over to the following national projects: ERB, ERBE, and Clouds and the Earth's Radiant Energy System (CERES).
- Based on what has been learned from pitfalls associated with this radiometer, SSEC has been inspired to establish a better climate benchmark with CLARREO and ARI (Absolute Radiance Interferometer).

Column 3:
- Early McIDAS included archive capability, made possible by inventive modification to slant track tape recorders (used in TV video replay) developed by Suomi and his son Eric, led to SSEC being the National GOES Archive for 25 years. McIDAS is going strong at 43 years of age (1973–2016) with hundreds of national and international users (operational prediction centers, aviation weather providers, researchers, NASA Kennedy Space Center, National Hurricane Center, and Storm Prediction Center).
- AVS visible imaging has led to GOES visible and IR imaging.
- AMVs continue to positively impact NWP.

Column 4:
- VAS experiment now incorporated into GOES Sounder series.
- 2002–future: High spectral resolution sounders on polar satellites. The *Suomi NPP* is the latest.

In summary, Suomi's vision has led to net-flux measurements of the planets; the use of *Explorer VII* radiation balance instruments with ERB, ERBE, and CERES; the incorporation of ATS imaging on GOES VIS and IR series and AMVs in operation; and the incorporation of the VAS experiment with the GOES Sounder series.

Passing the Baton

Suomi's 25 years of service as the founding director and leader of SSEC was successful by any measure. But no matter how successful the founders of an institution, they do not serve in perpetuity. It is often difficult for the organization to maintain its foundation and remain spirited and steadfast in its accomplishments after the founders die. There are a host of reasons for failure, but perhaps the most important is dilution of a leader's singular vision where committee actions put that vision in peril. Committee decisions that altered the visions of Apple Computer's Steve Jobs and architect Frank Lloyd Wright have challenged their legacies.

James Fisk, one of the notable leaders at Bell Labs—president from 1959 until his retirement in 1974—had a recipe for success of a scientific laboratory. Quoting Fisk,

> The best science and technology will only result from bringing and holding together the best minds . . . But to bring and hold such minds together takes a lot of doing. They must be located, attracted, and challenged. The purpose of the place must be broad enough to give them room and its management wise enough to give them freedom. The free play of creative minds is necessary to the individuals themselves, to the atmosphere they find stimulating, and to their achievement in the interest of business. (Mabon 1975, p. 3)

Suomi adhered to several of these tenets—especially his ability to attract top-level scientists and give them freedom. The scientific ventures that took place (and are still taking place) at SSEC allowed the scientists to expand their area of interest and career goals. And once a scientist or engineer or technician exhibited strength in contributing to a project, Suomi did not forget it. There was a sense of security in positions at SSEC. That is, if a project was nearing completion or if funding fell short of expectations, those valued employees were shifted to another project with funding. Bob Fox was an especially considerate leader, and he would keep in contact with all members of the SSEC team and let them know the funding situation. In a "soft money" institution like SSEC, where survival depends on continuity of support from outside organizations, this sense of security has always been impressive. It is certainly not the case in most other soft money institutions.

As Vern Suomi was nearing the end of his career and relinquishing directorship of the center, he exhibited a degree of peacefulness, realizing that a cadre of his protégés was poised to maintain continuity of science and engineering at the center. The situation was not unlike the one that occurred at the Mayo Clinic after Will and Charlie Mayo, the brothers who founded the clinic, retired and died soon thereafter. Harry Harwick, longtime business manager of the clinic, made the following statement on his retirement in 1950: "No one man, we knew, could ever replace Doctor Will's administrative genius, Doctor Charlie's incomparable human touch. This being so, we had to replace the best we could these great individual qualities with teamwork" (Harwick 1957).

And there was a team at SSEC that replaced Suomi and remained true to his singular vision. Despite the fact that neither a document nor a speech on this subject exists, it is not difficult to pen a statement close to Suomi's vision. In his style, the expression should be simple: *people and purpose*. The purpose would be advancement of science and engineering in service to society—not essentially different from Robert Millikan's view as the chair of the executive council at Caltech (Goodstein 1991). And he always realized that an organization's most valuable commodity is its people. The team of Suomi protégés and other excellent leaders within SSEC and CIMMS reached out and smoothly took the baton from Suomi's hand. Any visitor who worked at SSEC during Suomi's reign notices little difference in the operation and spirit of the center when he or she visits today.

Final Thoughts

Vern Suomi created an environment at SSEC that had many of the Mayo Clinic's and Bell Laboratories' attributes. His excitement about science was effervescent; he demanded much from his troops, but he also had the gift of encouragement. Suomi followed the style of his mentor Rossby by inviting a host of visitors from the United States and around the world to visit SSEC, thereby ensuring a dynamic, vibrant exchange of ideas. In turn, he encouraged SSEC scientists to spend time at other institutions, an activity that continues to this day. For those of us who worked with and alongside Verner Suomi, we remember his voice with varying intonations, his intense interest in a problem, and his Socratic questioning during a seminar. As we walk the hallways and look into the labs and offices at SSEC, we are reminded of our meetings and conversations with him at those places. And even if we were looking back at some aspect of our research during this meeting with Suomi, we were looking forward to the next challenge at the conversation's end.

APPENDIX A

Robert J. Parent's Vita

Birth: March 8, 1917, Menominee, MI

Education:

- Grade School/High School: Crivitz, WI
- University of Wisconsin–Madison
    - BS, Electrical Engineering, 1939
    - MS, Electrical Engineering, 1949
        - Thesis: A Video Frequency Phase Measuring Set

Military Service:

- Major, U.S. Army Signal Corps (1941–1946)
- Radar Instructor:
    - Harvard University (1942–1943)
    - MIT (1943–1945)
    - U.S. Army Artillery School, Fort Sill, OK (1945–1946)

Professional Experience:

- Engineer, General Electric Co., Schenectady, NY (1940–1941)
- Instructor–Professor, Department of Electrical Engineering, UW–Madison, 1946–1978
    - Assistant Professor, 1949
    - Associate Professor (Tenure Granted), 1953
    - Professor, 1958

Figure A.1. Robert J. Parent

- Director of Electrical Standards and Instrumentation Laboratory (College of Engineering, UW–Madison), 1953–1966
- Associate Director of Instrumentation Systems Center (College of Engineering, UW–Madison), 1966–1978
- Associate Director of Space Science and Engineering Center (UW–Madison), 1967–1973

Significant Contributions to Research at SSEC:

- In association with V. E. Suomi, development of radiometers aboard the following satellites: Vanguard, *Explorer VII*, *TIROS-3*, *TIROS-4*, and *TIROS-5*, ESSA and ITOS series
- In association with V. E. Suomi, development of the spin-scan cloud camera aboard *ATS-1*
- Ground-based data processing systems for TIROS
- Subminiature digital recorder for data accumulation and storage system for ESSA series of satellites

Key Publications (with V. E. Suomi):

- Parent, R. J., H. H. Miller, V. E. Suomi, and W. B. Swift, 1959: Instrumentation for a thermal radiation budget satellite. *Proc. National Electronics Conf.*, **15**, 34–46.

- Suomi, V. E., and R. J. Parent, 1960: A simple high capacity digital output data storage system for space experiments. *IRE Transactions, Fifth National Symp. on Space Electronics and Telemetry.*

Professional Societies:

- Instrument Society of America
- Institute of Electrical and Electronic Engineers
- Tau Beta Pi (Engineering Honor Society)
- Sigma Xi (Scientific Research Society)

Awards:

Arnold O. Beckman Award, Instrument Society of America Citation: For conceiving, developing, and guiding the construction of advanced instrument applications for satellite-borne measurement systems.

Death: June 23, 1978, Madison, WI

Memorial Service, UW–Madison, 11 September 1978

Memorial Statement: Bob Parent was noted for his interest in reducing theory to practice, and he was one of those rare individuals with the creativity to conceive complex equipment, the genius to design it, and the persistence to make it work. He will be remembered for is unfailing kindness and good humor.

APPENDIX B

Vignettes

Judi (Suomi) Maki

My earliest memory of my Uncle Vern is a trip to visit the Suomi family in Madison, Wisconsin, when I quite young, perhaps 7 years old. . . . I met a new friend, Ann Bryson, and giggled when my uncle talked about hearing the corn grow. I remember my dad commenting on how his brother always took apart more than he put together when he saw the shelves of "broken" stuff in his basement.

Over the years there were many times that my uncle and his family came to stay with us in Eveleth, Minnesota, as part of their summer vacations. Most of the Suomi family remained in the area including his sisters Anne, Edith, and Esther. . . . My dad was especially grateful to have time with him.

In later years, he and Paula enjoyed time at my family's cabin on Lake Vermillion where he experimented with his Styrofoam plate contraption (as he called it) measuring the amount of heat energy being released by the waves.

I was honored to invite my uncle to be the speaker at my high school graduation and thrilled that he followed my request to make it short. He spoke for only five minutes with an indelible message: Ask questions!

Uncle Vern and Auntie Paula helped my son through law school at UW. Vern said it was payback since my dad abandoned his dream of becoming a lawyer to put him through college. I do not ever recall any jealousy on my dad's part, only immense pride in everything Vern ever accomplished.

William Smith

I first met Professor Suomi in 1963 when I went to graduate school at the University of Wisconsin–Madison. Professor Suomi always made his classes interesting by skillfully using everyday analogies and demonstrating principles with simple experiments. When I went up to UW to meet with Prof. Suomi, there was a long line of students waiting to talk to him. Instead of waiting in line I walked down the hall where I met Prof. Lyle Horn and his PhD student Donald Johnson. They were also working on atmospheric energetics. After talking to them, Prof. Horn offered me an assistantship to continue my BS thesis work towards a MS degree under his guidance. I decided to accept his offer since I believed that I would receive much more individual attention from Professor Horn than Professor Suomi, since Professor Suomi already had an abundance of students . . . since I decided to work on the problem of retrieving atmospheric profiles from satellite spectral radiance measurements, Professor Horn suggested that I have Professor Suomi as my thesis advisor. Thus, for my PhD, I actually had two advisors, Professor Horn was my academic advisor and Professor Suomi was my thesis advisor. I was very close to both of these individuals.

Prof. Suomi and Dave Johnson had a very close relationship. Because of Professor Suomi's success in measuring the Earth's outgoing radiation from low Earth orbit and his innovation to use geostationary satellites for observing the Earth's weather, Dave Johnson relied heavily on Professor Suomi's judgment and strongly supported his research programs at the UW. I remember once when a proposal review committee recommended not to fund one of Suomi's proposals because they didn't think the proposed idea was sound. Dave Johnson overrode the committee recommendation stating something like "I don't care what the proposal says, we always get positive results from the research performed at the UW under Suomi's direction. We will fund this proposal."

Pierre Morel

I first met Professor Suomi in the summer of 1967 on the small island of Skepparholmen near Stockholm. Skepparholmen was the venue of the launching conference of the (proposed) Global Atmospheric Research

Program (GARP). Suffice it to say that GARP was the brainchild of a few visionary atmospheric scientists in the USA and a few others, like Academician Bugaev of the USSR Hydrometeorological Service and Professor Bert Bolin of the University of Stockholm. The idea of a cooperative scientific research program with innocuous, even beneficial, applications had captured the interest of political leaders in the USA and the Soviet Union and was strongly supported by both governments as a useful alternative to further confrontation.

Vern Suomi was the first who befriended me and he stood out with a unique ability to bring rambling scientific or other arguments back on track. He was very effective in delivering his trademark exhortation to "fish or cut bait." But for all his plain speaking, Vern was a very easy to get along with, basically a very honest and likable person, precisely the mix of solid-based scientific expertise combined with deep-seated intellectual integrity (a rare commodity indeed in modern science society).

Because of his scientific maturity, Vern Suomi was able to embrace *all aspects of a problem*, from conceptual understanding of the physical roots to the specific technical details of instrument design or field measurement campaigns (including of course considerable technical expertise in all things electronic), combined with the ability to extract appropriate support for his projects. He was the ultimate all-around investigator (some would say operator), an endangered breed in our days of sound bites, instant "breakthroughs" and short-lived "fame." In other words, he was a Renaissance man within an ever more compartmentalized world of narrow specialization. Vern was a creative thinker who pioneered many productive innovations but his main contribution [was] to observational meteorology; his pioneering role in the development of geostationary meteorological satellites would suffice to establish his stature as one of the most influential environmental scientists of the century.

Johannes Schmetz

I met Verner Suomi for the first time in the late 1980s when I was working at European Space Agency's (ESA) Space Operations Centre (ESOC) in the Meteosat Project where the first generation Meteosat satellites were operated. I had started working on calibration and the derivation of meteorological products just a few years earlier. We were very pleased about the invitation to visit the SSEC in Madison, which we admired as the center of gravity in satellite meteorology. It is obvious that I benefitted from the discussions with Prof. Suomi in many ways although the first visits and later visits until the early 1990s were quite short. I vividly remember that when I visited I was brought up to date on the ideas spinning around the center. My interest and work in satellite meteorology focused at that time on getting the first generation Meteosat's usefully calibrated and on deriving meteorological products for use in numerical weather prediction, notably the Atmospheric Motion Vectors (AMVs) which Vern Suomi had pioneered too. I have also special memories of the 'practical wisdom' that Vern Suomi often expressed in discussions. Let me quote one example which I tried to use as personal guidance as much as possible in projects and especially when I did set up a new team at EUMETSAT. At one of the visits to Madison I heard Vern Suomi say something like: 'I find good scientists and then I get out of their way.' I thought that is a very special approach that shows absolute confidence in his people. Many years later, I attended a well-known management school in Europe, the Saint Gallen Business School, and I was amazed to hear that they advised to act in very much the same way in order to make their organizations more effective and productive. Although I didn't visit SSEC often, and I only met Vern Suomi a few days in my life, he changed my perception of many things and I can say that he broadened my view and extended my horizon.

Stephen Cox

In the fall of 1967 we learned that Verner was to receive the Rossby Research Medal Award at the annual AMS meeting the following January. Tom Vonder Haar and I were working as research associates at the UW SSEC at the time. Since we were delivering papers at the AMS meeting, we made plans to travel to the meeting with Verner. When the day arrived to leave for the east coast, the Madison airport was fogged in. We waited all day, hoping the fog would clear. Verner was excited about receiving the award and anxious about the address he was expected to give. In the 5 years I had known him, I had not seen him in this anxious state.

After examining our options, we decided to drive to O'Hare which was supposed to clear before Madison, but it would be very close to make it in time for the flight.

We rented a car and then the issue of who should drive came up. It made sense that the "younger" eyes might best be able to see in this dense fog. And then there was the reaction time argument in favor of youth. But Verner, always uncomfortable with indecision, made the following argument.

He should drive because he was far sighted. As such, he could not see the light scattering fog droplets close to the car so they would not bother him as much. Tom and I scratched our heads but who were we to argue with our mentor. The outcome was a "white knuckle" ride to O'Hare, arriving barely in time for the flight and getting Verner to the AMS meeting to receive the award.

Henry Revercomb

(A letter to Vern Suomi while he was hospitalized in July 1995)

This is a heart-felt thank you for your nature as reflected in the Center you bore. I hope it is my privilege to thank you for a long time, as you have always done for those that worked with you.

SSEC and you have had a very positive impact on my life. The blend of objectivity and chaotic irrationality is enigmatic, but the undiluted dedication to understanding nature (and people, and fun, and exploration, and ego, and obfuscation, at times) is clearly unique. You have set an example, not perfect and God-like, but flawed and yet superhuman; unmatchable.

You engender the super energy and power that can be unleashed by genuinely inspired interest in solving important problems for mankind (and the challenge of it).

Frederick House

Dr. Suomi was a good listener and supported good ideas of his graduate students. On one occasion he visited our apartment at 408B Eagle Heights. He knocked on the door and Rose [my wife] answered and invited him inside. He saw my oldest son Fred (4 years old), picked him up, went outside and walked around in the sunshine. Fred didn't mind the attention at all. Dr. Suomi was a kind and gentle person.

James Rasmussen

I attended St. Olaf College in Northfield, Minnesota, for undergraduate, finished in 1958, and then served in the Air Force for three years. Then, graduate work at Colorado State on the water balance of the Colorado River Basin—under Herb Riehl I was the first student in the atmospheric sciences program when it arose out of the engineering department, and that's where I first met Vern Suomi when he came to visit and assist Herb Riehl with the setup of the department. I felt honored to be involved with the conversations of the department setup as the only student in the program at the time. When I finished I stayed on the CSU faculty along with Steve Cox and Tom Vonder Haar until 1972 and then went into federal service.

Here's a story that involves Vern. While I was working in Geneva in accordance with the GARP project, Suomi had a birthday celebration (not sure which age), I, Suomi and several other GARP committee members went to dinner (at the Versoix Restaurant in Geneva) and afterward Suomi and many in the group returned to our apartment where my wife Sonja had organized a birthday celebration for Suomi (cake and snacks). When Suomi arrived, he immediately sought interest in my children's activities (they were in the 3rd and 5th grade at the time). He discussed books with my daughter—and my son was building miniature towns inside matchboxes which Suomi took special interest in and even bought a town from him in the amount of 2 franks—this encouraged my son to continue his building—my wife and I were so excited and impressed with Suomi where at a large birthday celebration for himself he took more interest in our children and their activities.

Douglas Sargeant

I first met Suomi just south of Cedar Rapids, Iowa, at the Collins Radio Company field research facility, ~30 miles north of Iowa City; I was trying to decide what to do—go to grad school or begin working, I was near finished with my undergraduate degree in physics at [University of Iowa] in Iowa City.

George Ludwig, who I met briefly at Iowa City, suggested I consider attending UW–Madison graduate school for engineering.

When I arrived in Madison, I met with several faculty in both physics and meteorology. I was so impressed with Suomi because he was so enthusiastic about his work. Suomi was an expert lecturer, extremely convincing man, but also a great listener. I told Suomi about my undergraduate research at Iowa City in connection with the Collins Radio Company on atmospheric scatter and microwave propagation. He took an immediate interest in my work. He helped me prepare a proposal, in his name, that was funded on tropospheric scatter and microwave propagation over the horizon. This became my graduate work.

APPENDIX C

Mentorship

The top part of the tree (see Figure C.1) has been extracted from Lewis (1992). Rossby was Suomi's original doctoral advisor at University of Chicago, but after Rossby's departure for the International Institute of Meteorology at the University of Stockholm in the early 1950s, Horace Byers became Suomi's doctoral advisor. In addition to the doctoral students listed under Suomi, a group of young scientists who came under his influence as postdoctoral students are listed as mentees: Smith, Sromovsky, Revercomb, and Menzel. There are 26 PhDs listed under Suomi's name, and he also supervised 56 master's degree (MS) students (not listed).

The following statements from several protégés give the reader some idea of Suomi's style of training young scientists. Years of primary mentorship are placed in parentheses after the protégés name.

Frederick House (early 1960s)

As Dr. Suomi's first student in satellites, I was processing *Explorer VII* data using a circular slide rule. . . . It was a period of excitement and real innovation to solve problems as they came up, especially after satellites were in orbit. Many times I prepared observational results from satellites that were used at various meetings he attended. In the summer of 1963 . . . Dr. Suomi

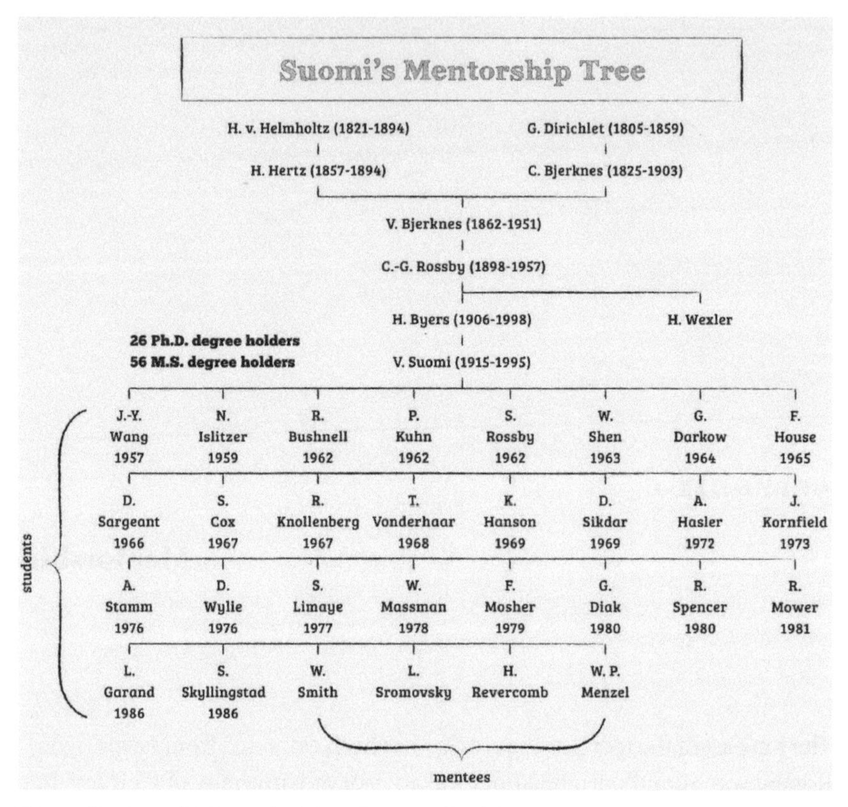

Figure C.1. Suomi's Mentorship Tree. Courtesy Sarah Witman.

was in Europe and then traveled to San Francisco for the meeting. We met [in San Francisco] and discussed slides of results based on TIROS IV measurements. He selected slides for his presentation the next day. At this point I felt he trusted the quality of my work.

Stephen Cox (1960s)

Verner was most definitely a mentor to me!! He was my MS advisor from 1962 to 1964, my PhD advisor from 1964 to 1967, and my postdoctoral sponsor from 1967 to 1969. In addition to supervising my academic and research endeavors, he showed, by example, how enthusiasm and commitment can be contagious.

Verner typically assigned his students to a research associate or senior graduate student to help him/her learn the ropes. I was told to report to Peter Kuhn, a PhD candidate studying under Verner and working on the same Weather Bureau contract that supported my assistantship. During my first year of study Verner would meet with me individually to plan my courses and in a group of his advisees to learn about his research interests, his contract and grant research obligations and what other students were working on.

Verner was a strong believer in giving his students a lot of freedom in conceiving and pursuing their graduate program. Some might call it the "sink or swim" philosophy. He offered guidance if he thought you were going too far off track, but gave you enough flexibility to learn from your mistakes. He would offer advice when requested and at periodic progress report meetings, but he did not micromanage.

Verner's attention to his students increased significantly as they approached completion of their thesis. He preferred receiving a complete document, not partial installments. Upon reading the thesis he would have the student come to his office where he would render his opinion of the work. Verner did not mince words. His response would range from "this is rubbish" to "here are some things that you need to address." There were often multiple iterations before the thesis was given to the student's committee.

Thomas Vonder Haar (mid-1960s to early 1970s)

By the middle 60s, Vern had a keen appreciation of professorships and academia. He wanted to be called "professor" and not "doctor." I remember he told me at the convocation at the Fieldhouse (winter of 1967 I believe), "the PhDs stand up first, before the medical doctors and veterinarians." Vern's method of mentorship was "sink or swim." He created opportunities.

Vern was very attentive on the final versions of dissertations. He always had time for those. He would often require students to do more work on their dissertations than they expected. He came back to me when all my radiation budget stuff was done. It was a nice complete dissertation, my committee liked it, Vern liked it. But then he said at the "11th hour," "but you should be able to get the solar constant." I worked another month and had an appendix in my thesis where I tried to get the solar constant. It came out to be 1400 Watts per meter squared but I had an error budget. He wasn't going to sign off until I included that calculation. He was great in the beginning, very attentive and rigorous at the end. And I recall he was that way with a number of other students including Steve Cox and Bob Knollenberg. He kept the bar very high. It was an interesting bi-modal approach to students. He was good at getting them inspired and then kind of ignored them. And then he would be there at the end.

Paul Menzel (mid- to late 1970s)

There were three of us from the physics department at UW (PhDs in physics) who joined SSEC. Larry Sromovsky was the first and he had more work than he could handle and he told Vern there's another guy named Hank Revercomb who could help. The two of them had more work than they could handle and then they brought me in. Suomi was very positive about that—he wanted physicists. My job interview with Vern went something like this: VS: "Do you know any meteorology?" And I figured "here it goes." VS said: "That's great, I'll teach you all you need to know." He didn't want me to have any of these pre-conceived notions.

Vern thought that there were enough resources in the building [SSEC] and created a team of people that were there to serve the student. He didn't feel like he had to be there to take care of the student, there are enough resources without involving him, except perhaps for an initial discussion with the student who says, "this is an idea I have, what do you think about it." A lot of times I don't even think he knew who his students were. I have stories to corroborate that. There would be a little social engagement and somebody would be talking to Suomi and Suomi would say, "who are you working with," and the student would say "I'm your student." Suomi: "Oh!" He figured there is an exchange of ideas and there is a fertile ground here at SSEC, if you can't figure out how to make this work for you, then it is probably not right that you get a degree from me.

Vern would grab whoever was handy and invite us to the B&B with a distinguished visitor. We were always made to feel a part of the important group that included the visiting "hall of famer."

William Smith

One personal story I have. Actually, Professor Suomi was not my major advisor, I mean not my major professor . . . that was Professor Lyle Horn. Suomi was my primary PhD thesis advisor. . . . The story I remember most is actually on the Sunday the day before my PhD defense. I had given Professor Suomi my thesis many weeks earlier, but never heard anything from him. So I thought I better call him and make sure he knew about my defense being the next day. And when I talked to him it seemed like he didn't realize that I had my defense the next day. But he never let on that was the case; instead he said 'Bill come out to the house.' Give me three hours and come on out to the house and we'll talk about your thesis—which I did. He obviously read my thesis well because we had a very good discussion of it at his home. And then near the end of our discussion Paula his wife invited me to stay for dinner which I graciously accepted. And I got to know Suomi's wife for the very first time then. And I remember thinking during dinner that old adage that behind every great man there's a great woman because Paula was certainly that. Well, the story doesn't end there. The next day my oral defense took place and it went pretty well except for one question that Professor Suomi asked me . . . he says 'why can you see further with infrared remote sensing looking down from a satellite than looking up from the ground?' And I really didn't know the answer to that. But Professor Suomi went on to explain that infrared remote sensing is like looking through a series of glass panes—each atmospheric level being a different pane. And he says the panes get dirtier going from the top down due to a phenomena called the pressure broadening of absorption lines. But he says when you're looking up from the ground, your looking through the dirtiest panes first so you can't see very far. When you're looking down from the top, you're looking through the cleanest ones first and therefore you can see further . . . I did appreciate how that simple question and the answer would impact my professional career from then until this very day [November 2, 2009].

APPENDIX D

Suomi's People: List of Coworkers, Protégés, and Colleagues

Charles Anderson. Professor of Meteorology at UW–Madison and first Executive Director of SSEC (1965–1967)

Richard Anthes. A UW–Madison doctoral student of Professor Don Johnson who became director of NCAR and pioneering scientist for the COSMIC Project

John Benson. Software Engineer and McIDAS mainstay

Fred Best. Mechanical engineer who improved the radiometric calibration of infrared sensors

Francis Bretherton. Geophysical science theoretician, President of University Corporation for Atmospheric Research before replacing Suomi as Director of SSEC in 1991

Reid Bryson. Founder of UW–Madison's Meteorology Department (1947), who made Suomi his first hire, and where both men had a life-long collegial relationship despite differing opinions on climate and weather puzzles

Horace Byers. Doctoral student of C.-G. Rossby at MIT, central to the Cadet Program at University of Chicago, and later served as Suomi's doctoral dissertation advisor after Rossby's departure for Stockholm University

Gene Buchholtz. SSEC electronics technician who made important contributions to many of the early instruments

Ralph Dedecker. SSEC engineer who contributed software programs for aircraft test bed instruments flown on ER2

George Diak. Suomi doctoral student—expert in use of skin temperature observations from satellite

Robert Dombroski. SSEC mechanical engineer who worked on the Pioneer Venus radiometers

Vicki Epps. Suomi's secretary at SSEC—she ran the show

Michael Ference. Physics professor at University of Chicago who was "the best teacher I [V.S.] ever had"

Robert Fox. AFIT student at UW–Madison, later received PhD at UW, led AF programs in remote sensing before becoming fourth Executive Director of SSEC (1980–1995)

Theodore "Ted" Fujita. Colleague from University of Chicago who collaborated and competed with Suomi to produce the first color movies of the Earth. He became famous for his five scales of tornado severity [Fujita scale (F-scale)]

Terry Gregory. Administrative assistant at SSEC who handled interactions with the media

Thomas Haig. Air Force Colonel and NRO scientific leader who supported the Suomi–Parent radiometer work and became third Executive Director of SSEC (1970–1980)

Kirby Hanson. USWB–SSEC liaison (late 1960s), second Executive Director of SSEC (1968–1970) and PhD under Suomi

Christopher "Kit" Hayden. NOAA scientist who moved to Madison to develop the VAS applications; he later became head of the NESDIS Systems Design and Applications Branch at SSEC

Robert Herbsleb. Electronics technician who built and fine-tuned the instruments designed for investigating the atmospheres of neighboring planets

William Hibbard. Mathematician/computer scientist who developed 3-D software that enabled model forecast visualization

Lyle Horn. Professor of Meteorology at UW–Madison whose doctoral students worked closely with SSEC scientists

Frederick House. Suomi doctoral student—processed *Explorer VII* radiation data, TIROS data, and made accurate estimates of the Earth–atmosphere heat budget

David Johnson. Pioneering director of NESS (NESDIS) who supported many of Suomi's initiatives and strengthened the ties between NOAA and SSEC with CIMSS as an outgrowth

Donald Johnson. Professor of Meteorology at UW–Madison, trusted colleague of Suomi and mentor to several high-ranking leaders in U.S. meteorology

Graeme Kelly. ANMRC meteorologist brought to SSEC by Bill Smith to design data analysis schemes for VAS observations

Terry Kelly. Television meteorologist who introduced the public to McIDAS products

Robert Krauss. Engineer/scientist who coauthored the Suomi–Parent paper that introduced the idea of the spin-scan camera

John LeMarshall. ANMRC meteorologist brought to SSEC by Bill Smith to work on temperature and wind retrieval from satellites

John Lewis. A member of Bill Smith's NESDIS group who was brought to SSEC to assimilate VAS soundings into dynamic models (with help from John Derber, Jim Purser, Lee Panetta, Graham Mills, and Andy van Tuyl)

Sanjay Limaye. Suomi doctoral student who used cloud tracking techniques to capture circulations on Venus

Jim Maynard. SSEC engineer who managed the in-house hardware production of McIDAS

Paul Menzel. Physicist hired by Suomi in 1975 to join the VAS effort; he subsequently became the first occupant of the Suomi Distinguished Professorship within the Department of Atmospheric and Oceanic Sciences at UW–Madison

Graham Mills. Operational weather forecaster from ANMRC who was brought to SSEC by Vern Suomi and John Lewis to work on VAS data

Pierre Morel. French physicist who led major international programs in space science, initiated the European Meteosat geostationary satellite system, and served with Suomi on the Joint Organizing Committee of GARP

Frederick Mosher. Suomi doctoral student who improved cloud tracking of atmospheric motion vectors and introduced McIDAS to the NSSFC (now SPC)

Dave Nelson. Doctoral student in electrical engineering who served as Robert Parent's assistant in radiometer design and magnetic storage

Homer Newell. Head of NASA's Office of Space Sciences in the 1960s who facilitated addition of the spin-scan camera on *ATS-1*

Robert Oehlkers. Electronics technician—50-year veteran of SSEC projects

Robert Parent. Ingenious electrical and mechanical engineering professor at UW–Madison who was central to success of early satellite meteorology projects at UW and SSEC

Jean Phillips. SSEC librarian, archivist, nationally-recognized historian of satellite meteorology and former chair of AMS's History of Science Committee

Jim Purser. Theoretical physicist (Met Office, UK) brought to SSEC by John Lewis and Kit Hayden to use his recursive filter analysis technique on satellite data

Henry (Hank) Revercomb. Physicist hired by Suomi in 1973 to work on the ERBE and VAS programs; he subsequently became Director of SSEC

Evan Richards. Engineer who managed the technicians associated with building the instruments for investigating the atmospheres of neighboring planets

Herbert Riehl. The "father of tropical meteorology" who served on Suomi's doctoral committee at University of Chicago

John Roberts. Suomi's accountant, the celebrated "bean counter," who balanced SSEC's books

Ragnar Rollefson. UW–Madison professor of physics who was a contemporary of Suomi

Carl-Gustaf Rossby. Among the 20th century's greatest theoreticians in geophysical science who exerted great influence on Suomi while he attended the University of Chicago

Tony Schreiner. A Suomi protégé who processed VAS data and developed satellite products used in operations

Leo Skille. Early electronics technician who became the SSEC building manager and overseer of SSEC projects

Bill Smith. A Lyle Horn–Vern Suomi doctoral student who became a NESS lead scientist, then co-principal investigator with Suomi on geo-sounding, and brought his NESDIS group to SSEC and helped establish CIMSS

Larry Sromovsky. Physicist hired by Suomi in 1972 to lead the ERBE and VAS efforts and became his primary sanity check on new ideas; he still works on datasets collected during the NASA planetary explorations

Robert Sutton. Machinist at SSEC whose innovative work was central to success of many SSEC projects

Anard Suomi. Vern's older brother, his hero and guide, who became the breadwinner for the large Suomi family after their father's death

Eric Suomi. Youngest son of Vern Suomi and SSEC engineer who pioneered the video cassette archive of GOES data

Paula Suomi. Vern Suomi's wife and life-partner who raised their three children and taught grade school in Madison

Roger Thompson. Lead engineer at Santa Barbara Research Systems who worked on the spin-scan camera

Bill Togstad. Operational NWS forecaster brought to SSEC as a member of Bill Smith's NESDIS group with responsibility to prove the value of VAS soundings

Louis Uccellini. Protégé of Professor Don Johnson, lead scientist at NASA for the VAS program, and current Director of the National Weather Service

Christopher Velden. SSEC scientist whose work with cloud-tracked winds has served the world community of meteorologists

Thomas Vonder Haar. Suomi doctoral student—first to find the now-standard value of the Earth's albedo (0.30) using satellite data

Dee Wade. Head of SSEC's computer data center

Gary Wade. A member of Bill Smith's NESDIS group who excelled at satellite data manipulation and construction of beautiful graphics in support of research papers and journal articles

Ken Walker. Electronics technician who contributed to the building of the early net-flux sensors

Tony Wendricks. Draftsman of many early instrument plans and raconteur of the many Suomi/SSEC stories

Harry Wexler. Rossby doctoral student at MIT who became Chief of Research for the USWB and a major contributor to Antarctic meteorology and central figure in the IGY

Robert White. Pioneer in studies of general circulation at MIT in the 1950s who went on to become the last Chief of the USWB, supervising Suomi's work at the bureau in the early 1960s

John Wiley. UW–Madison professor of physics and contemporary of Suomi who became Chancellor of UW–Madison in 2001

Robert Wollersheim. Lead engineer on many early SSEC projects, established Wollersheim Winery in Sauk City after retirement

J. T. Young. Meteorologist and computer scientist who guided early McIDAS applications

APPENDIX E

Suomi's Witticisms and Aphorisms

Remembered by Chris Velden:

- Processing satellite data is like taking a drink from a fire hydrant
- Don't just ask "why," ask "why not"

Remembered by Reid Bryson:

- When you run out of ideas, buy equipment
- An instrument should have one tube or less
- The important thing in science is knowing what you can set equal to zero and get away with

Remembered by Steve Cox:

- The worst enemy of a good idea is a better idea
- The KISS principle (Keep It Simple, Stupid)

References

Achtor, T., T. Rink, T. Whittaker, D. Parker, and D. Santek, 2008: McIDAS-V: A powerful data analysis and visualization tool for multi and hyperspectral environmental satellite data. *Atmospheric and Environmental Remote Sensing Data Processing and Utilization IV: Readiness for GEOSS II*, M. D. Goldberg et al., Eds., International Society for Optics and Photonics (SPIE Proceedings, Vol. 7085), 708509, https://doi.org/10.1117/12.795223.

Aldrich, L. B., 1922: The reflecting power of clouds. *Annals of the Astrophysical Observatory of the Smithsonian Institution*, Vol. 4, Smithsonian Institution, 375–382.

Andersson, E., J. Pailleux, J.-N. Thépaut, J.R. Eyre, A.P. McNally, G.A. Kelly and P. Courtier, 1994: Use of cloud-cleared radiances in three/four-dimensional variational data assimilation. *Q. J. Royal Meteorol. Soc.*, **120**, 627–653.

Armstrong, B., 1989/90: A gift for simplicity. Research Sampler, 27-31. Madison, WI, Graduate School, University of Wisconsin–Madison: https://search.library.wisc.edu/catalog/999558209802121

Barrett, E. W., and V. E. Suomi, 1949: Preliminary report on temperature measurement by sonic means. *J. Meteor.*, **6**, 273–276, https://doi.org/10.1175/1520-0469(1949)006<0273:PROTMB>2.0.CO;2.

Bell, E. T., 1937: *Men of Mathematics*. Simon and Schuster, 592 pp.

Bergeron, T., 1981: Synoptic meteorology: An historical review. *Weather and Weather Maps*, G. H. Liljequist, Ed., Brikhäuser, 443–473, https://doi.org/10.1007/978-3-0348-5148-0_2.

Bergthorsson, P., and Döös, B., 1955: Numerical weather map analysis. *Tellus*, 7, 329–340.

Bolin, B., 1959: *The Atmosphere and Sea in Motion: Scientific Contributions to the Rossby Memorial Volume*. Rockefeller University Press, 509 pp.

Boltzmann, L., 1884: Ableitung des Stefan'schen Gesetzes, betreffend die Abhängigkeit der Wärmestrahlung von der Temperatur aus der electromagnetischen Lichttheorie. *Ann. Phys.*, **258**, 291–294, https://doi.org/10.1002/andp.18842580616.

Boyden, M., V. Oliver, and J. Holland, 1942: An example of a field forecast. University of Chicago Misc. Rep. 6, Part II, Institute of Meteorology, 22–43.

Bourke, W., et al, 2004: History of NWP in Australia-1970 to the present. Proc. 16th BMRC Modelling Workshop, Melbourne, Australia, 6-9 December 2004. (Available from National Library of Australia: https://catalogue.nla.gov.au/Record/3354061)

Businger, J., 2005: Reflections on boundary layer problems of the last 50 years. *Bound.-Layer Meteor.*, **116**, 161–173, https://doi.org/10.1007/s10546-004-4712-1.

Byers, H. R., 1970: Recollections of the war years. *Bull. Amer. Meteor. Soc.*, **51**, 214–217, https://doi.org/10.1175/1520-0477(1970)051<0214:ROTWY>2.0.CO;2.

Cadet Program, 1941: Nine-month course in meteorology. Institute of Meteorology, University of Chicago, 5 pp.

Carissimo, B. C., A. H. Oort, and T. H. Vonder Haar, 1985: Estimating the meridional energy transports in the atmosphere and ocean. *J. Phys. Oceanogr.*, **15**, 82–91, https://doi.org/10.1175/1520-0485(1985)015<0082:ETMETI>2.0.CO;2.

Chandrasekhar, S., 1950: *Radiative Transfer*. Oxford University Press, 393 pp.

Charney, J. G., 1948: On the scale of the atmospheric motions. *Geofys. Publ.*, **17** (2), 3–17.

———. R. Fjortøft, and J. von Neumann, 1950: Numerical integration of the barotropic vorticity equation. *Tellus*, **2**, 237–254, https://doi.org/10.1111/j.2153-3490.1950.tb00336.x.

Crepeau, J., 2007: Josef Stefan: His life and legacy in the thermal sciences. *Exp. Therm. Fluid Sci.*, **31**, 795–803, https://doi.org/10.1016/j.expthermflusci.2006.08.005.

Cressman, G.P., 1959: An operational objective analysis system. *Mon. Wea. Rev.*, **87**, 367–374.

Day, D., 1999: Bergen West: Or how four Scandinavian geophysicists found a home in the New World. *Hist.-Meereskd. Jahrb.*, **6**, 69–82.

Dee, D. P., M. Balmaseda, G. Balsam, R. Engelen, A. J. Simmons, and J. N. Thepaut, 2014: Toward a consistent reanalysis of the climate system. *Bull. Am. Meteor. Soc.*, **95**, 1235–1248.

Derber, J. C., and Wu, SW.S., 1998: The use of TOVS cloud-cleared radiances in the NCEP SSI analysis system. *Mon. Wea. Rev.*, **126**, 2287–2299.

Dulong, P. L., and A. T. Petit, 1817: Des Recherches sur la Mesure des Températures et sur les Lois de la communication de la chaleur. *Ann. Chim. Phys.*, **7**, 225–264.

Dyson, F. J., 1988: *Infinite in All Directions*. Harper & Row, 321 pp.

Ellis, J. S., T. Vonder Haar, S. Levitus, and A. Oort, 1978: The annual variation in the global heat balance of the Earth. *J. Geophys. Res.*, **83**, 1958–1962, https://doi.org/10.1029/JC083iC04p01958.

Eyre, J., 2007: Progress achieved on assimilation of satellite data in NWP over the last 30 years. ECMWF Seminar on Recent Development in Use of Satellite Observations in NWP, 3-7 September 2007, 53 pp.

Eyre, J. R., 1997: Variational assimilation of remotely-sensed observations of the atmosphere. *J, Met. Soc. Japan*, **75**, 1B, 331–338.

———. G. A. Kelly, A. P. McNally, E. Andersson and A. Persson, 1993: Assimilation of TOVS radiance information through one-dimensional variational analysis. *Q. J. Royal Meteorol. Soc.*, **119**, 1427–1463.

Ference, M., and H. Snodgrass, 1942: A mobile weather unit. University of Chicago Misc. Rep. 6, Part I, Institute of Meteorology, 21 pp.

———. and V. Suomi, n. d.: Survey and analysis of sonic devices for measuring true air speed. University of Chicago Special Collections (Dept. of Meteor. Records), University of Chicago Archives (box 2, folder 7).

Fleming, J. R., 1998: *Historical Perspectives on Climate Change*. Oxford University Press, 208 pp.

———. 2007: *The Callendar Effect. The Life and Work of Guy Stewart Callendar* (1898–1964). Amer. Meteor. Soc., 155 pp.

———. 2016: *Inventing Atmospheric Science: Bjerknes, Rossby, Wexler, and the Foundations of Modern Meteorology*. MIT Press, 296 pp.

Fox, R. J., 2009: Remembering Verner Edward Suomi: The father of satellite meteorology. Oral history session moderated by R. Fox, 50th Anniversary of Explorer VII, Monona Terrace Community and Convention Center, Madison, WI.

Fritz, S., 1949: The albedo of the planet Earth and of clouds. *J. Meteor.*, **6**, 277–282, https://doi.org/10.1175/1520-0469(1949)006<0277:TAOTPE>2.0.CO;2.

Fujita, T. T., 1992: *Memoirs of an Effort to Unlock the Mystery of Severe Storms: During the 50 Years 1942–1992*. University of Chicago, 298 pp.

Fujita, T. T. and Bohan, W., 1967: Detailed Views of Mesoscale Cloud Patterns filmed from ATS-1 Pictures. 16 mm film, silent, color. University of Chicago, Department of Geophysical Sciences, Satellite and Mesometeorology Research Project. Images from Verner E. Suomi, University of Wisconsin–Madison, Space Science and Engineering Center. Sponsored by National Aeronautics and Space Administration (NASA) and Environmental Science Services Administration (ESSA).

Gandin, L., 1965: *Objective Analysis of Meteorological Fields*. Israel Program for Scientific Translations, 242 pp.

Gavaghan, H., 1998: *Something New Under the Sun*. Springer, 300 pp.

Glickman, T., Ed., 2000: *Glossary of Meteorology*. 2nd ed. Amer. Meteor. Soc., 855 pp., http://glossary.ametsoc.org/.

Goerss, J. S., C. S. Velden, and J. D. Hawkins, 1998: The impact of multispectral GOES-8 wind information on Atlantic tropical cyclone track forecasts in 1995. Part II: NOGAPS forecasts. *Mon. Wea. Rev.*, **126**, 1219–1227.

Goode, P. R., J. Qui, V. Yurchyshyn, J. Hickey, M.-C. Chu, E. Kolbe, C. T. Brown, and S. E. Koonin, 2001: Earthshine observations of the Earth's reflectance. *Geophys. Res. Lett.*, **28**, 1671–1674, https://doi.org/10.1029/2000GL012580.

Goodstein, J. R., 1991: *Millikan's School, A History of the California Institute of Technology*. W. W. Norton & Co., 317 pp.

Goody, R. M., 1964: *Atmospheric Radiation I: Theoretical Basis*. Oxford Monographs on Meteorology, Clarendon Press, 436 pp.

———. and J. C. G. Walker, 1972: *Atmospheres*. Foundations in Earth Science Series, Prentice-Hall, 150 pp.

Gregory, T., 1998: Verner Edward Suomi (6 December 1915–30 July 1995). *Proc. Amer. Philos. Soc.*, **142**, 500–509, http://www.jstor.org/stable/3152256.

———. T. Achtor, J. Phillips, and T. Haig, 2005: From heat budget of the Earth to interferometry: Suomi and Parent's legacy. *Third Presidential History Symp.*, San Diego, CA, Amer. Meteor. Soc., 1.5, https://ams.confex.com/ams/Annual2005/techprogram/paper_85745.htm.

Guymer, L., 1969: Estimation of 1000–500 mb thickness patterns from satellite pictures of convective areas. *Proc. WMO Inter-regional Seminar on the Interpretation of Meteorological Satellite Data*, Melbourne, VIC, Australia, Bureau of Meteorology, 51–55.

Haig, T. O., 1977: The McIDAS system. *Proc. Workshop on Interactive Video Displays for Atmospheric Studies*, SSEC Publ. 77.08.11, Madison, WI, University of Wisconsin–Madison, 171–178.

Hartmann, D. L., 1994: *Global Physical Climatology*. Academic Press, 411 pp.

Harwick, H. J., 1957: *Forty-Four Years with the Mayo Clinic: 1908–1952*. Whiting, 50 pp.

Harwood, J. J., 2001: Michael Ference. *Memorial Tributes: National Academy of Engineering*, Vol. 9, National Academies Press, 76–80.

Hayden, C. M., P. N. Belov, A. I. Bourtzev, E. P. Dombkovskaja, A. V. Karpov, S. I. Solovjev, A. B. Uspenski, and P. M. Weintreb, 1979: Remote soundings of temperature and moisture. *Quantitative Meteorological Data from Satellites*, WMO Tech Note 166, 102 pp., https://library.wmo.int/pmb_ged/wmo_531.pdf.

Houghton, H. G., 1954: On the annual heat balance of the northern hemisphere. *J. Meteor.*, **11**, 1–9, https://doi.org/10.1175/1520-0469(1954)011<0001:OTAHBO>2.0.CO;2.

Houghton, J. T., F. W. Taylor, and C. D. Rodgers, 1984: *Remote Sounding of Atmospheres*. Cambridge University Press, 343 pp.

House, F. B., 1965: The radiation balance of the Earth from a satellite. PhD thesis, Dept. of Meteorology, University of Wisconsin–Madison, 69 pp.

———. A. Gruber, G. E. Hunt, and A. T. Mecherikunnel, 1986: History of satellite missions and measurements of the Earth radiation budget (1957–1984). *Rev. Geophys.*, **24**, 357–377, https://doi.org/10.1029/RG024i002p00357.

Hunt, G. E., R. Kandel, and A. T. Mecherikunnel, 1986: A history of presatellite investigations of the Earth's radiation budget. *Rev. Geophys.*, **24**, 351–356, https://doi.org/10.1029/RG024i002p00351.

IGY, 1961: Earth's thermal radiation balance: Preliminary results from Explorer VII. IGY Bulletin, *Eos, Trans. Amer. Geophys. Union*, **42**, 467–474, https://doi.org/10.1029/TR042i004p00467.

Johnston, H. I., 2001: Oliver Reynolds Wulf. *Biographical Memoirs*, Vol. 79, National Academies Press, **79**, 396–411.

Jones, W. W., and Staff Members, 1943: Weather-map construction and forecasting in the westerlies from single-station aerological data. University Chicago Misc. Rep. 7, Institute of Meteorology, 35 pp.

Kaplan, L. D., 1959: Inference of atmospheric structure from remote radiation measurements. *J. Opt. Soc. Amer.*, **49**, 1004–1007, https://doi.org/10.1364/JOSA.49.001004.

Kashara, A., J. Tribbia, and W. Washington, 1987: Oral history interview with Aksel Wiin-Nielsen. American Meteorological Society Oral History Project, UCAR, 62 pp., https://opensky.ucar.edu/islandora/object/archives%3A7658.

Kelly, G., 1997: Influence of observations on the operational ECMWF system. WMO Bulletin, 46, 4, 336–342.

Kelly, G., G. Mills, and W. Smith, 1978: Impact of Nimbus-6 temperature soundings on Australian regional forecasts. *Bull. Amer. Meteor. Soc.*, **59**, 393–405.

Key, J.R., D. Santek, C.S. Velden, et al., 2003: Cloud-drift and water vapor winds in the polar regions from MODIS. *IEEE Trans. Geosci. Rem. Sens.* **41**, 482–492.

Kiehl, J. and W. Washington, 1987: Oral history interview with Julius London. American Meteorological Society Oral History Project, UCAR, 26 pp., https://opensky.ucar.edu/islandora/object/archives%3A8306.

King, J. I. F., 1958: The radiative heat transfer of planet Earth. *Scientific Uses of Earth Satellites*. 2nd rev. ed., J. A. Van Allen, Ed. University of Michigan Press, 133–136.

Krauss, E. B., and J. S. Turner, 1967: A one-dimensional model of the seasonal thermocline II. The general theory and its consequences. *Tellus*, **19**, 98–106, https://doi.org/10.1111/j.2153-3490.1967.tb01462.x.

Lazzara, M. A., and Coauthors, 1999: The Man Computer Interactive Data Access System: 25 years of interactive processing. *Bull. Amer. Meteor. Soc.*, **80**, 271–284, https://doi.org/10.1175/1520-0477(1999)080<0271:TMCIDA>2.0.CO;2.

Lazzara, M.A., R. Dworak, D.A. Santek, et al., 2014: High-latitude atmospheric motion vectors from composite satellite data. *J. Appl. Meteor. Climatol.*, **53**, 534-547.

Lettau, H. H., and B. Davidson, Eds., 1957: *Instrumentation and Data Evaluation*. Vol. 1, *Exploring the Atmosphere's First Mile*, Pergamon Press, 376 pp.

Lewis, J. M., 1992: Carl-Gustaf Rossby: A study in mentorship. *Bull. Amer. Meteor. Soc.*, **73**, 1425–1438, https://doi.org/10.1175/1520-0477(1992)073<1425:CGRASI>2.0.CO;2.

———. 1996: C.-G. Rossby: Geostrophic adjustment as an outgrowth of modeling the Gulf Stream. *Bull. Amer. Meteor. Soc.*, **77**, 2711–2728, https://doi.org/10.1175/1520-0477(1996)077<2711:CGRGAA>2.0.CO;2.

———. 1998: Clarifying the dynamics of the general circulation: Phillips's 1956 experiment. *Bull. Amer. Meteor. Soc.*, **79**, 39–60, https://doi.org/10.1175/1520-0477(1998)079<0039:CTDOTG>2.0.CO;2.

———. 2008: Smagorinsky's GFDL: Building the team. *Bull. Amer. Meteor. Soc.*, **89**, 1339–1353, https://doi.org/10.1175/2008BAMS2599.1.

———. and J. C. Derber, 1985: The use of adjoint equations to solve a variational adjustment problem with advective constraints. *Tellus*, **37A**, 309–322, https://doi.org/10.1111/j.1600-0870.1985.tb00430.x.

———. C. Hayden, C. Velden, T. Stewart, R. Lord, S. Goldenberg, and S. Aberson, 1987(a): The use of VAS winds and temperatures as input to barotropic hurricane track forecasting. Preprints 16th Conference on Hurricanes and Tropical Prediction, Houston, TX.

———. and S. Lakshmivarahan, 2008: Sasaki's pivotal contribution: Calculus of variations applied to weather map analysis. *Mon. Wea. Rev.*, **136**, 3553–3567.

———. S. Lakshmivarahan, and S. K. Dhall, 2006: *Dynamic Data Assimilation: A Least Squares Approach*. Cambridge University Press, 654 pp.

———. D. W. Martin, R. M. Rabin, and H. Moosmüller, 2010: Suomi: Pragmatic visionary. *Bull. Amer. Meteor. Soc.*, **91**, 559–577, https://doi.org/10.1175/2009BAMS2897.1.

Lewis, J., A. Van Tuyl, and C. Velden, 1987(b): A dynamical method for building continuity into the deep layer mean wind. *Mon. Wea. Rev.*, **115**, 885–893.

Lilly, D. K., 1968: Models of the cloud-topped mixed layers under a strong inversion. *Quart. J. Roy. Meteor. Soc.*, **94**, 292–309, https://doi.org/10.1002/qj.49709440106.

Loeb, N. G., B. A. Wielicki, D. R. Doelling, G. L. Smith, D. F. Keyes, S. Kato, N. Manalo-Smith, and T. Wong, 2009: Toward optimal closure of the Earth's top-of-atmosphere radiation budget. *J. Climate*, **22**, 748–766, https://doi.org/10.1175/ 2008JCLI2637.1.

London, J., 1957: A study of the atmospheric heat balance. Final Rep., Contract AF 19(122)-165, Dept. of Meteorology and Oceanography, New York University, 99 pp.

Lorenc, A., 1981: A global three-dimensional multivariate statistical analysis scheme. *Mon. Wea. Rev.*, **109**, 701–721, https://doi.org/10.1175/1520-0493(1981)109<0701:AGTDMS>2.0.CO;2.

Mabon, P. C., 1975: *Mission Communications: The Story of Bell Laboratories*. Bell Telephone Laboratories, Inc., 198 pp.

Malkevich, M. S., V. M. Pokras, and L. I. Yurkova, 1963: Measurements of radiation balance on the satellite Explorer VII. *Planet. Space Sci.*, **11**, 839–865, https://doi.org/10.1016/0032-0633(63)90196-X.

Mayo, C. W., 1968: *Mayo: The Story of My Family and My Career*. Doubleday and Co., 351 pp.

Menzel, W. P., 1995: Verner Suomi (1915–1995). *Eos, Trans. Amer. Geophys. Union*, **76**, 454–455, https://doi.org/10.1029/95EO00285.

Menzel, W.P., 2001: Cloud tracking with satellite imagery: From the pioneering work of Ted Fujita to the present. *Bull. Amer. Meteor. Soc.*, **82**, 1, 33–47.

———. W. L. Smith, and L. D. Herman, 1981: Visible infrared spin-scan radiometer atmospheric sounder radiometric calibration: An inflight evaluation from intercomparisons with HIRS and radiosonde measurements. *Appl. Opt.*, **20**, 3641–3644, https://doi.org/10.1364/AO.20.003641.

———. ———. G. S. Wade, L. D. Herman, and C. M. Hayden, 1983a: Atmospheric soundings from a geostationary satellite. *Appl. Opt.*, **22**, 2686–2689, https://doi.org/10.1364/AO.22.002686.

———. ———. and T. R. Stewart, 1983b: Improved cloud motion wind vector and altitude assignment using VAS. *J. Climate Appl. Meteor.*, **22**, 377–384, https://doi.org/10.1175/1520-0450(1983)022<0377:ICMWVA>2.0.CO;2.

Mertz, R., 1971: Interview with Joseph Smagorinsky, 19 May 1971. Computer Oral History Collection, National Museum of American History, Smithsonian Institution, 100 pp.

NASA, 1985: VAS Demonstration: VISSR Atmospheric Sounder description and final report. NASA Reference Publ. 1151, 170 pp., https://ntrs.nasa.gov/archive/nasa/casi.ntrs.nasa.gov/19860004398.pdf.

Neufeld, M. J., 2007: *Von Braun: Dreamer of Space, Engineer of War*. Vintage Books, 587 pp.

Ohring, G., 1979: Impact of satellite temperature sounding data on weather forecasts. *Bull. Amer. Meteor. Soc.*, **10**, 1142–1147.

Oort, A. H., and T. H. Vonder Haar, 1976: On the observed annual cycle in the ocean–atmosphere heat balance over the Northern Hemisphere. *J. Phys. Oceanogr.*, **6**, 781–800, https://doi.org/10.1175/1520-0485(1976)006<0781:OTOACI>2.0.CO;2.

Petersen, R., and R. Aune, 2007: An objective nowcasting tool that optimizes the impact of satellite derived sounder products in very-short-range forecasts. *11th Symp. on Integrated Observing and Assimilation Systems for the Atmosphere, Oceans, and Land Surface*, San Antonio, TX, Amer. Meteor. Soc., 3.17, https://ams.confex.com/ams/87ANNUAL/webprogram/Paper117341.html.

Phillips, J., 2014: The world according to GATE (and GARP). *Through the Atmosphere*, Summer/Fall 2014, Space Science and Engineering Center, University of Wisconsin–Madison, 5–7, http://www.ssec.wisc.edu/through-the-atmos/ttasummer2014/.

Phillips, N., 1956: The general circulation of the atmosphere: A numerical experiment. *Quart. J. Roy. Meteor. Soc.*, **82**, 123–164, https://doi.org/10.1002/qj.49708235202.

Rabson, D., 1998: It happened here: The invisible ally. UCAR Staff Notes, October 1998, http://www.ucar.edu/communications/staffnotes/9810/here.html.

Ramanathan, V., 2008: Why is the Earth's albedo 29% and was it always 29%? *iLEAPS Newsletter*, No. 5, IGBP, University of Helsinki, Helsinki, Finland, 18–20.

Revercomb, H. E., H. Buijs, H. B. Howell, D. D. LaPorte, W. L. Smith, and L. A. Sromovsky, 1988: Radiometric calibration of IR Fourier transform spectrometers: Solution to a problem with the High-resolution Interferometer Sounder. *Appl. Opt.*, **27**, 3210–3218, https://doi.org/10.1364/AO.27.003210.

Riordan, M., and L. Hoddeson, 1997: *Crystal Fire: The Birth of the Information Age*. W. W. Norton & Co., 352 pp.

Rossby, C.-G., 1943: Preliminary report on the activities of the University Meteorological Committee. University of Chicago Archives, 13 pp.

———. 1952: Letter to Jule Charney, dated April 16, 1952. Jule Gregory Charney Papers, MC184, Institute Archives and Special Collections, Massachusetts Institute of Technology, Cambridge, MA.

Russell, H. N., 1916: On the albedo of the planets and their satellites. *Astrophys. J.*, **43**, 173–196.

Rutherford, I., 1972: Data assimilation by statistical interpolation of forecast error fields. *J. Atmos. Sci.*, **29**, 809–815, https://doi.org/10.1175/1520-0469(1972)029<0809:DABSIO>2.0.CO;2.

Sanders, F., and R. W. Burpee, 1968: Experiments in barotropic hurricane track forecasting. *J. Appl. Meteor.*, **7**, 313–323.

Schatz, R., 2005: *From Finland with Love* (*Suomesta rakkaudella*). J. Kniga, 131 pp.

Seaman, R. S., R. L. Falconer, and J. Brown, 1977: Application of a variational blending technique to numerical analysis in the Australian region. *Aust. Meteor. Mag.*, **25**, 3–23.

Simmons, A. J., and A. Hollingsworth, A., 2002: Some aspects of the improvement in skill of numerical weather prediction. *Quart. J. Roy. Meteor. Soc.*, **128**, 647–677, https://doi.org/10.1256/003590002321042135.

———. R. Mureau, and T. Petroliagis, 1995: Error growth and predictability estimates for the ECMWF forecasting system. *Quart. J. Roy. Meteor. Soc.*, **121**, 1739–1771, https://doi.org/10.1002/qj.49712152711.

Smagorinsky, J., 1978: History and progress. *The Global Weather Experiment—Perspectives and Implementation and Exploitation*. Rep. of FGGE Advisory Panel, National Academy of Sciences, 4–12.

———. and N. A. Phillips, 1978: Scientific problems of the global experiment. *The Global Weather Experiment—Perspectives and Implementation and Exploitation*. Rep. of FGGE Advisory Panel, National Academy of Sciences, 13–21.

Smith, E., and D. Phillips, 1972: WINDCO: An interactive system for obtaining accurate cloud motions from geostationary satellite spin scan camera pictures. Meteorological measurements from satellite platforms: Annual scientific report on NASA contract NAS 5-11542, 1970–1971, SSEC Publ.72.02.M1, University of Wisconsin–Madison, 1–51.

Smith, W. L.: 1985: Satellites. In Handbook of Applied Meteorology, D. D. Houghton, ed. John Wiley and Sons, 380–472.

Smith, W. L., and Coauthors, 1981: First sounding results from VAS-D. *Bull. Amer. Meteor. Soc.*, **62**, 232–236.

———. V. E. Suomi, F.-X. Zhou, and W. P. Menzel, 1982: Nowcasting applications of geostationary satellite atmospheric sounding data. *Nowcasting*, K. H. Browning, Ed., Academic Press, 123–135.

Sommer, B. W., 2008: *Hard Work and Good Deal: The Civilian Conservation Corps in Minnesota*. Minnesota Historical Society Press, 205 pp.

Sromovsky, L. A., 2015: Planetary science research at SSEC: History and current focus. Space Science and Engineering Center Tech. Rep., 14 pp.

———. J. R. Anderson, F. A. Best, J. P. Boyle, C. A. Sisko, and V. E. Suomi, 1999a: The Skin-Layer Ocean Heat Flux Instrument (SOHFI). Part I: Design and laboratory characterization. *J. Atmos. Oceanic Technol.*, **16**, 1224–1238, https://doi.org/10.1175/1520-0426(1999)016<1224:TSLOHF>2.0.CO;2.

———. ———. ———. ———. ———. and ———. 1999b: The Skin-Layer Ocean Heat Flux Instrument (SOHFI). Part II: Field measurements of surface heat flux and solar irradiance. *J. Atmos. Oceanic Technol.*, **16**, 1239–1254, https://doi.org/10.1175/1520-0426(1999)016<1239:TSLOHF>2.0.CO;2.

SSEC, 1973: Boundary Layer Instrumentation System. SSEC Publication No.73.04.B1, 11 pp., http://library.ssec.wisc.edu/instrumentation/Documents/BLIS/BLIS_007.

———. 1995: Verner E. Suomi, 1916–1995: A man for all seasons. SSEC Publ. 98.03.S1, University of Wisconsin–Madison, 30 pp., http://library.ssec.wisc.edu/SuomiWebsite/SuomiImages/scanned%20documents/Man_For_All_Seasons.pdf.

Štefan, J., 1879: Über die Beziehung zwischen der Wärmestrahlung und der Temperatur. *Sitzungsber. Bayer. Akad. Wiss., Math.-Naturwiss. Kl.*, **79**, 391–428.

Stephens, G. L., D. O'Brien, P. J. Webster, P. Pilewski, S. Kato and J. l. Li, 2015: The albedo of Earth. Rev. *Geophys.*, **53**, 141–163, https://doi.org/10.1002/2014RG000449.

Suomi, V. E., 1953: The heat budget over a cornfield. PhD dissertation, Dept. of Meteorology, University Chicago, 49 pp.

———. 1957: Sonic anemometer—University of Wisconsin. *Exploring the Atmosphere's First Mile*, H. Lettau and B. Davidson, Eds., Pergamon Press, 256–266.

———. 1961: The Thermal Radiation Balance Experiment on board Explorer VII. Juno II Summary Project Report, Volume I: Explorer VII Satellite, NASA Tech. Note D-608, 273–305.

———. and R. J. Parent, 1964: A proposal to NASA for spin scan camera for ATS-C. SSEC Publ. 64.00.S1, University of Wisconsin–Madison, 12 pp.

———. and R. J. Krauss, 1978: The spin scan camera system: Geostationary meteorological satellite workhorse for a decade. *Opt. Eng.*, **17**, 170106, https://doi.org/10.1117/12.7972172.

———. and J. Lewis, 1990: Oral history of Verner Suomi, 18 May 1990. 1 tape (not transcribed).

———. M. Franssila, and N. F. Islitzer, 1954: An improved net-radiation instrument. *J. Meteor.*, **11**, 276–282, https://doi.org/10.1175/1520-0469(1954)011<0276:AINRI>2.0.CO;2.

———. R. J. Fox, S. S. Limaye, and W. L. Smith, 1983: McIDAS III: A Modern Interactive Data Access and Analysis System. *J. Climate Appl. Meteor.*, **22**, 766–778, https://doi.org/10.1175/1520-0450(1983)022<0766:MIAMID>2.0.CO;2.

———. Da. Johnson, Do. Johnson, G. Kutzbach, and W. Smith, 1994: Oral history of Verner Suomi, 20 March, 20 April, 14 May 1994. American Meteorological Society Oral History Project, UCAR, 3 tapes (2 transcribed), http://library.ssec.wisc.edu/SuomiWebsite/Oral_history.html.

———. L. A. Sromovsky, and J. R. Anderson, 1996: Measuring ocean-atmosphere heat flux with a new in-situ sensor. Preprints, *8th Conf. on Air–Sea Interactions and Conf. on the Global Ocean–Atmosphere–Land System (GOALS)*, Atlanta, GA, Amer. Meteor. Soc., 38–42.

Tabor, H., 1956: An instrument for measuring absorptivities for solar radiation. *J. Scientific Instr.* **33**, 356–358.

Thompson, P. D., 1983: A history of numerical weather prediction in the United States. *Bull. Amer. Meteor. Soc.*, **84**, 755–769.

Trenberth, K. E., J. T. Fasullo, and J. T. Kiehl, 2009: Earth's global energy budget. *Bull. Amer. Meteor. Soc.*, **90**, 311–323, https://doi.org/10.1175/2008BAMS2634.1.

Troup, A. J., and N. A. Streten, 1972: Satellite-observed Southern Hemisphere cloud vortices in relation to conventional observations. *J. Appl. Meteor.*, **11**, 909–917, https://doi.org/10.1175/1520-0450(1972)011<0909:SOSHCV>2.0.CO;2.

Tyndall, J., 1868: *Heat Considered as a Mode of Motion*. D. Appleton & Company, 541 pp.

United Nations, 1961: John F. Kennedy address at U.N. General Assembly, 25 September 1961. (Available from John F. Kennedy Presidential Library and Museum: https://www.jfklibrary.org/Asset-Viewer/DOPIN64xJUGRKgdHJ9NfgQ.aspx).

University of Chicago, Division of the Physical Sciences, 1950: Announcements. 50, 8, University of Chicago: http://pi.lib.uchicago.edu/1001/cat/bib/4693820.

Velden, C., J. Daniels, D. Stettner, D. Santek, J. Key, J. Dunion, K. Holmlund, G. Dengel, W. Bresky, and P. Menzel, 2005: Recent innovations in deriving tropospheric winds from meteorological satellites. *Bull. Amer. Meteor. Soc.*, **86**, 205–223.

Velden, C.S., T.L. Olander, and S. Wanzong, 1998: The impact of multispectral GOES-8 wind information on Atlantic tropical cyclone track forecasts in 1995. Part I: Dataset methodology, description, and case analysis. *Mon. Wea. Rev.*, **126**, 1202–1218.

Very, F. W., 1912: The Earth's albedo. *Astron. Nachr.*, **196**, 269–290.

Vonder Haar, T. H., 1968: Variations of the Earth's radiation budget. PhD dissertation, Dept. of Meteorology, University of Wisconsin–Madison, 118 pp.

———. and V. E. Suomi, 1969: Satellite observations of the Earth's radiation budget. *Science*, **163**, 667–668, https://doi.org/10.1126/science.163.3868.667.

———. and ———. 1971: Measurements of the Earth's radiation budget from satellites during a five-year period. Part I: Extended time and space means. *J. Atmos. Sci.*, **28**, 305–314, https://doi.org/10.1175/1520-0469(1971)028<0305:MOTERB>2.0.CO;2.

———. and A. Oort, 1973: New estimate of annual poleward energy transport by Northern Hemisphere oceans. *J. Phys. Oceanogr.*, **3**, 169–172, https://doi.org/10.1175/1520-0485(1973)003<0169:NEOAPE>2.0.CO;2.

Walters, R., 1952: *Weather Training in the AAF* (1937–1945). USAF Historical Study 56, U.S. Air Force, 234 pp., http://www.afhra.af.mil/Portals/16/documents/Studies/51-100/AFD-090529-110.pdf.

Wanner, W., A. H. Strahler, B. Hu, P. Lewis, J.-P. Muller, X. Li, C. L. Barker Schaaf, and M. J. Barnsley, 1997: Global retrieval of bidirectional reflectance and albedo over land from EOS MODIS and MISR data: Theory and algorithm. *J. Geophys. Res.*, **102**, 172143–172161, https://doi.org/10.1029/96JD03295.

Weinman, J., 2003: Interview of James A. Weinman on 20 December 2003. University of Wisconsin Oral History Program, 2 tapes, http://digital.library.wisc.edu/1793/70297.

Wexler, H., 1954: Observing the weather from a satellite vehicle. *J. Br. Interplanet. Soc.*, **13**, 269–276.

———. 1960: Satellites and meteorology. *WMO Bull.*, **9** (1), 2–7, https://library.wmo.int/pmb_ged/bulletin_9-1_en.pdf.

———. and M. Neiburger, 1961: Weather satellites. *Sci. Amer.*, **205**, 80–94.

Whittaker, T. M., and R. Dedecker, 1977: Processing and display of satellite and conventional meteorological data for the classroom. Interactive video displays for atmospheric studies, SSEC Publication No.77.08.I1, University of Wisconsin–Madison, 11–24, http://library.ssec.wisc.edu/edocs/whitaker_mcidas_demo_1977.pdf.

Wielicki, B. A., T. M. Wong, N. Loeb, P. Minnis, K. Priestley, and R. Kandel, 2005: Changes in Earth's albedo measured by satellite. *Science*, **308**, 825, https://doi.org/10.1126/science.1106484.

Wiin-Nielsen, A., 1991: The birth of numerical weather prediction. *Tellus*, **43AB**, 36–52.

Wunch, C., 1997: Memoir of Henry Stommel. *Biographical Memoirs*, Vol. 72, National Academies Press, 321–350.

Young, J. T., and R. J. Fox, 1994: Future plans for McIDAS workstations. *Fourth Workshop on Meteorology Operational Systems*, Reading, United Kingdom, ECMWF, 240–247.

Zillman, J. W., and P. G. Price, 1972: On the thermal structure of mature Southern Ocean cyclones. *Aust. Meteor. Mag.*, **20**, 34–38.

Zuckerman, H., 1977: *Scientific Elite: Nobel Laureates in the United States*. Free Press, 335 pp.

Index